AF772646

African Economic Transformation in the Digital Age

Editors

Mammo Muchie and Angathevar Baskaran

AFRICA WORLD PRESS
TRENTON | LONDON | CAPE TOWN | NAIROBI | ADDIS ABABA | ASMARA | IBADAN | NEW DELHI

AFRICA WORLD PRESS
541 West Ingham Avenue | Suite B
Trenton, New Jersey 08638

Copyright © 2017

All rights reserved. No part of this publication may be reproduced, stored in a retrieval system or transmitted in any form or by any means electronic, mechanical, photocopying, recording or otherwise without the prior written permission of the publisher.

Book design: Dawid Kahts
Cover art: Ananthy Baskaran

Library of Congress Cataloging-in-Publication Data

Names: Muchie, Mammo, editor. | Baskaran, Angathevar, editor.
Title: African economic transformation in the digital age / Editors, Mammo Muchie & Angathevar Baskaran.
Description: Trenton : AFRICA WORLD PRESS, 2017. | Includes bibliographical references and index.
Identifiers: LCCN 2017008501| ISBN 9781569025574 (hbk : alk. paper) | ISBN 9781569025581 (pbk : alk. paper)
Subjects: LCSH: Information technology--Economic aspects--Africa. | Digital communications--Economic aspects--Africa. | Economic development--Effect of technological innovations on--Africa.
Classification: LCC HC800.Z9 A345 2017 | DDC 338.9600285--dc23
LC record available at https://lccn.loc.gov/2017008501

African Economic Transformation in the Digital Age

We dedicate this book to

Ananthy Baskaran
&
Adey & Mikael with Fikre, Eradu and Hasset Getachew Muchie

And all the African children to live with hope making a better world with intellectual distinction and intelligence

Contents

Contents

Preface:
The Digital Age is Upon the World

Mammo Muchie & Angathevar Baskaran

We are now entering what is called the 4th industrial revolution. The first industrial revolution was driven by water and steam power to mechanize production. Electricity to create mass production powered the second industrial revolution. Electronics, information and communication technologies to automate production defined the third industrial revolution. Now a Fourth Industrial Revolution is emerging blending the physical, ICTs and biological worlds with a variety of technologies such as new materials, alternative energy storage, solar energy, quantum computing, robotics, artificial intelligence, the internet, nanotechnology, 3D printing and other moving and building upon the digital revolution that has been occurring since the middle of the last century.

The world is now going through a technological revolution that requires all humans to find shared approaches to respond to and deal with the changes. The fourth industrial revolution is driven by exponential technologies that produce unusual disruptive innovations compared with previous industrial revolutions.

The Fourth industrial revolution is evolving exponentially by changing almost every industry in every country. Humanity cannot go living as in the previous industrial revolution times of the first, second and third. There is a breadth and depth in the changes that can herald transformation of entire systems of production, management, and governance. These changes are marked with the scale, scope and complexity never seen before without also changing how people live, work and relate with one another.

The fourth industrial revolution stands on the shoulders of the digital age with a velocity, scope and system impact that is not easy to predict and know in advance. The fourth industrial revolution is driven by exponential technology. A new ecosystem is needed that can bring together all the arrangements to moderate and control the exponential technology bringing waves of innovation to create and destroy with production in every sector in the world as it is going right now.

The challenging question is what will be the impact of the fourth industrial revolution, exponential technologies and the universalisation of the digital age to transform Africa environmentally, socially, politically and economically by strengthening rather than diluting Pan-African integration? As Africa lacks infrastructure, electricity, how can it utlise the fourth industrial revolution and exponential technologies to get the continent to embark on the development path to leapfrog and catch up with the rest of the world? As Africa is still in the mineral and agriculture stage exporting minerals and agrarian goods and services, how can Africa use exponential technologies of the fourth industrial revolution to side-step the first, second and third industrial revolutions. This is a real dilemma of entering the digitial exponential technological revolutionary age while remaining still a predominantly agriculture lead economy.

There is a real challenge for Africa to examine how industrialisation with digitial and exponential technology driving can be facilitated and promoted. The varied contributions in this book bring out the critical problem of Africa's transformation in the digital age.

There is a need to do more research and bring out how Africa can use its resources by also emplying exponential technologies to arrive at a stage of sustainable and integrated development without any delay.

We recommend strongly that readers engage with this book which the contributors used creative scholarship to address an important problem of how new technology that disrupts need all in the world to come tgether and address how to make it beneficil rather than harmful.

We explore whether novel exponential technologies move African transformation forward. They can and the critical issue is learning how to manage the transformation dynamics so that people can do, gain, earn and learn more by transforming the way of thinking, doing and relating with the consumers and producers and with each other and the rest of the world.

Reference

Klaus Schwab, The Fourth Industrail Revolution, Portfold Penguin, 2016

Acknowledgements

We would like to thank the DST/NRF SARChI (Innovation and Development) at the Tshwane University of Technology, Pretoria, South Africa for initiating the Science, Technology, Innovation Studies series of books in collaboration with African World Press.

We recognize and appreciate the support given by Dr. Jideofor Adibe, Adonis & Abbey Publishers Ltd, London, United Kingdom, which enabled publication of this book.

We also would like to thank the African Journal of Science, Technology, Innovation and Development (AJSTID) and extend our fullest appreciation and thanks to all the contributing authors.

Our sincere thanks to Ananthy Baskaran, the cover designer, and Jan B. Mwesigwa (www.janmwesigwa.com) our initial layout artist for their hard work and dedication to the book project.

Finally, we express our deepest appreciation and gratitude to Mr. Kassahun Checole and Mr. Girma A. Demeke, Africa World Press, Inc. & The Red Sea Press, Inc. for their cooperation and support during the publication process. We also thank all the publications staff at Africa World Press, Inc. & The Red Sea Press, Inc.

The DST/NRF Research Chair at Tshwane University of Technology and Africa World Press will promote the science, technology, innovation studies book series applying the principle to promote innovation and knowledge on and for Africa.

Chapter 1
Introduction
African Economic Transformation in the Digital Age

Mammo Muchie & Angathevar Baskaran

Information and communication technologies or ICTs[1] are considered as bringing about a major paradigm shift by having brought far-reaching changes in the way interactions are taking place among peoples, economies, societies and communities in the world today. In other words, they have spearheaded the dawn of the 'Digital Age' where we have now vast masses of people across the globe becoming e-people and fewer and fewer people that have not gone electronic yet. The number of those who go e- is growing rapidly, and those who have not yet become computer un-literate appear to dwindle over time. That e- people are populating the world as electronic herds everywhere in the world making communication instantaneous is a reality that is with us now. It raises big opportunities as well as challenges. It is remarkable how recent political changes in nearly all parts of the world have been attributed to the ease of communication that ICTs provided among those who were able to coordinate protests. It looks everything that used to be done with human expenditure of energy has been changed with the application of ICTs in all walks of life from the military

1 Throughout this book (except few places) we use 'ICT' to mean information and communications technologies instead of 'ICTs' for convenience.

using drones to facilitating information on how to pick up passports and when to pay ones taxes. Humans are all going e- or failure to go e- is likely to make them fall behind and lose the opportunity to play an active part in day-to-day life-world activities and possibilities.

Today there is massive digital data in the virtual world. The real world has libraries that are physical. The virtual world is full of data that circulates with ease and speed across the world. The world is awash with a data revolution. As the ICT revolution globalises the world; equally important, the ICT penetration in Africa is growing especially mobile telephones, but a comprehensive ICT infrastructure is still waiting to be laid down. What is critical also is how this ICT infrastructure is going to get established and spread. There is a need to coordinate the local, national, regional and Africa wide levels of ICT programmes and founding a robust infrastructure that endures. It appears that coordinating an Africa wide drive and led ICT infrastructure is not easy to pull off. Whilst the importance of ICT for Africa's economic transformation is highly appreciated, there is a gap between the intention and the actual promotion of an ICT infrastructure across Africa. NEPAD and AU can play a significant role if the current Governments are able to learn to frame the economic transformation of Africa by taking an African perspective.

A major challenge in Africa is also the continued governance problems and crises. ICT can be used to make both governments' public institutions and the governance that makes government express relevance and ability to make citizens be satisfied with its services. Accountability by Government by making sure through the help of ICTs citizens participate effectively to use Government services to improve their well being is very essential. There is an urgent demand to make African Governments become responsive and show equitable and just policy to prioritise service delivery over all other considerations. The ability to coordinate from local to regional governments will be facilitated by the application of nationwide diffusion of ICTs.

Real transformation comes when the welfare of the people in Africa becomes a number one priority from the local to the state, regional and Africa levels. Real transformation comes when peoples' lives are changed. The ICT4D is critical to bring about such a change, not by itself alone, but also by including other very relevant dimensions.

The chapters in this book address the important problem of how to build an ICT infrastructure for African economic transformation. They do not put a closure but rather open this important research area so that rigorous scholarship feeds the policy learning needed to construct a comprehensive ICT4D technological infrastructure that can connect Africa reliably for a smart, inclusive and sustainable development.

The post cold-war era dawned an on-going process of globalisation where the global economy has been anchored increasingly on knowledge and becoming the "learning economy" (Lundvall, 1992). Knowledge and learning are now widely regarded as the key to sustained development and growth of economies as well as societies. As information and communication technologies (ICT) has become one of the main driving forces of globalisation, it has been increasingly perceived as a major factor determining the comparative advantage of nations and the competitiveness of their economies because of its potential to bring about socio-economic transformation. Like the first machine-driven industrial revolution, the ICT revolution has generated fundamental changes in the socio-economic life of people and nations across the world. Indeed, the pace of globalisation processes of socio-economic activities, witnessed since the early 1990s, has been largely propelled by technological change in ICT.

Unlike other technologies, the ICT is unique mainly in two aspects. First, it affects almost every country on the face of the earth and almost every aspect of life – economic, social, and cultural. And therefore the dynamics of technological change in ICT is more global than national or regional. Second, unlike other technologies, ICT as an enabler technology having significant potential to build both social and technological capabilities in developing countries. In recent years, it has already led to significant socio-economic transformation in both developed and developing countries. It is truly a global phenomenon particularly in terms of its potential applications (socio-economic), reach (market), attraction and utility to ordinary people across the world, associated rate of technological change, and opportunities for states – whether small or big, and industrialised or developing. On the other hand, the emergence of international governance system (intellectual property rights protecting information service providers, electronic commerce that needs greater network security from unwanted intrusion) that shapes the global expansion of ICT markets has created new challenges for national governments, particularly in the developing world.

One of the central issues examined in this book is whether the perception of ICT as a tool for all round socio-economic development is being reflected by the ground level developments/ realities or it is just a misplaced and over hyped 'faith' in a technology that is never going to deliver what it promises. The chapters covered in this book attempt to do this. They have mostly focused on the new innovation on the information and communications technologies (ICTs) sector, mobile phone and mobile money in Africa.

Some new innovative approaches in the ICT areas are being tried on both improving the supply side and also opening improved access on the demand side. On the supply side most of the improvements are driven by technological innovation, and on the demand side the challenge is in making economics not stand in the way to diffuse these new ICT, mobile phone and mobile money services to reach the rural areas of the numerous African economies and countries.

One of the novelest attempts on the supply side is the attempt to test solar-powered and water resistant ICT laptops. It has been reported in Ghana WeWi, that a Canadian Telecom company tried its water resistant and solar driven laptop in 2013. Similar developments are expected to be diffused in different African countries to have their water-resistant and solar powered laptops.

ICT infrastructure building for improving internet penetration in Africa is said to be growing. Some of the ICT infrastructures are the following: (a) Submarine cables that include Main One, WACS, ACE, Seacom, Glo-1, (b) Cable system that connects African countries with each other and the rest of the world, (c) Growth in under sea capacity to link with different parts of the world, (d) Creating data traffic re-routing of African countries to various parts of the world, (e) Broadband and wireless networks; (f) rolling out 3G and 4G, (g) Expansion of mobile broadband and integrating cloud computing with African ICT infrastructure. The supply side appears to be growing and new innovative ICT infrastructures are appearing in the African market.

On the demand side, the cost of bandwidth has to decrease and connectivity has to improve both in quality and durability. It is not only building new laptops that are solar-powered and water–resistant on the supply side that matter as innovations, there is also an important demand that needs to be met and that is to go beyond this and attempt to reach and access the bottom of the pyramid in the rural areas by creating mobile phones that are solar powered and water–resistant and other hurdles as well.

There are also efforts to create local content that have been tried, such as local languages (e.g. Swahili in Kenya). Microsoft and Huawei have developed mobile phones suitable for the average consumer in African markets by focusing on lowering the price and by creating A 420-hour stand by-mode battery life. Some of the user consumer compliant developments include local information uploading onto Google Maps; IROKOVtv for showing African TV shows on line; Snapplify for showcasing works for African publishers; Kleek for featuring African music; Ekitabu for users to purchase African books using M-Pesa, which is used widely in Kenya. These linkages of creating technologies and realising efficient service provisions are changing the economic landscape into an ICT led dynamic economic process.

ICT infrastructure for E-Learning in Africa is also evolving not just as creating more computers but also impacting on all economic sectors from schools to health, agriculture, construction and a host of other activities. The African Management Initiative has, for example, created MOOC as an online course on management. Technology innovation hubs as an incubation space are also developing by building local technology services. Various social problems as traffic congestion are being addressed with ICT infrastructure as part of the solution.

There are of course all these opportunities, but there are also challenges related to regulatory environment deficits, the dominance of few fixed line operators and a range of weaknesses that require an integration of e-Governance with facilitative regulatory conditions to promote an integrated and comprehensive ICT infrastructure, mobile phone and mobile money and other services on a predictable foundation and direction.

The system of innovation perspective emphasizes the identification in any given economic setting the interactions of the significant social-economic variables, and the dynamic co-evolution of institutions and technologies that result in imparting key development features and dynamics to a given national economy. We shall see in the case studies we have put together in this monograph how policies on ICT in different African countries contribute to imparting developmental features, effects and dynamics to these economies or the opposite as the case may be.

The main challenge faced by many African countries is how to integrate ICT as innovation-knowledge within specific national innovation systems, and also examine conversely how the national system of innovation diffuses ICT knowledge-innovations. The main objective is thus to undertake a double conceptualisation: the development of the concept of innovation systems stimulated and spearheaded by ICT and conversely how the existing innovation systems facilitate the creation, absorption and diffusion of ICT. This is one of the major issues examined in this book by the contributing research scholars. The book discusses the following topics.

- ICTs and economic transformation in Africa: a synthesis
- Innovation system approach towards economic transformation in the Digital Age
- Building the Information Society for us all
- Quantifying open source software development
- Research in African e-local governance: outcome assessment framework

- Policy implications of the relationship between ICT access and usage and well-being
- Assessing ICT impact on development and economic transformation in the rural areas
- The impact of the ICT policy on the printing and publishing industry
- An assessment of application software market
- Link between ICT investments, growth and jobs creation
- Impact of cellular phones on technical and allocative efficiencies of commercial taxi and fishing boat firms
- Impact of ICT investment, ownership and use in the Cassava value chain
- Conceptualizing accountability in ICT4D
- Sustainable broadband connectivity model for rural areas

In Chapter 2 Olu Ajakaiye and Samuel Wangwe present a synthesis of ICTs and economic transformation in Africa. In Chapter 3, Baskaran and Muchie discuss the application of innovation system approach towards economic transformation in the Digital Age.

The Chapter 4 illustrates neatly the broad spectrum of the late Christopher Freeman's vision on the way scientific and technological change, limited here to the so-called new information and communication technologies (ICTs) which were on the verge of impacting the global economy, would affect society. This chapter written in 1995 jointly with Luc Soete explores how the information society (IS) was a society in the process of formation. Starting from the perspective of the IS as a "learning society" important issues of how education and training, who controls information, the role of telecommunications in expanding services and a host of other changes have been identified and explored. The challenges to the organisation of work, on the structure of the firm, and the impact on job creation and destruction are also discussed. Both the opportunities of expanding ICTs and the possible negative consequences on regional cohesion, the provision of health services, the media and culture, as well as democracy are also addressed.

Michael Gastrow and Saahier Parker in Chapter 5 examine the issue of quantifying open source software development. In the context of increasing globalisation and the emergence of knowledge-based economies, the promotion of open source software (OSS) has received increased attention internationally. There exists a broad consensus that OSS has an important role to play in the development strategies of emerging economies, and in many countries, including South Africa, this has been manifested in policies supporting domestic open source software development. However, the current policy has been developed in the context of a paucity of reliable quantitative data describing open source software development in South Africa. This chapter therefore uses national Research and Development Survey data to provide a quantitative overview of open source software development in South Africa, with a view to informing future research directions in this area. The findings are supported by interviews with product development executives from four of South Africa's major software development firms. Key findings include a recent increase in expenditure on open source software development, and the identification of private sector clusters of activity in Gauteng and the Western Cape. Identified areas for potential further research include regional-level business-sector surveys, a focus on micro-level OSS developers, increased monitoring and evaluation of government OSS policies, and closer examination of the interaction between non-profit OSS advocates and government institutions.

In Chapter 6, Timothy Mwololo Waema, Winnie Mitullah and Edith Ader argue that the potential of ICT to contribute to good governance has been known for a long time and has been the subject of many articles and reports, but very little concrete empirical evidence of the effects of ICT on governance, and how these effects should be evaluated, exist. The situation is even worse when we examine ICT and governance local governments. First, drawing upon available literature on e-Government and e-Governance, this chapter proposes a research framework for assessing effects of e-Government projects on good governance. In the process, first, it adopts UNDP's nine characteristics of good governance which represent an adequate operational definition of good governance before focusing on three key applications of e-Government. Second, drawing on outcome monitoring and evaluation, the paper proposes a research framework for assessing the e-Governance outcomes as well as a set of good governance outcome indicators. The chapter illustrates the usefulness of the framework by providing the generic outcome indicators developed by LOG-IN Africa and outlining a study that was carried out in Kenya following the same framework. Third, suggesting that the proposed framework is a good starting point for investigating e-Governance outcomes, the paper invites researchers to test the framework and create the required knowledge in this area.

In Chapter 7 Timothy Mwololo Waema & Obadia Okind examine the relationship between information and communication technologies (ICTs) and well-being or poverty which has remained unclear and previous research has resulted in divergent conclusions. In Kenya, poverty reduction is unlikely to be attained through ICT applications when this understanding is blurred. This chapter is a synthesis of a study which sought to shed more light on this relationship. The study was carried out using a longitudinal methodology. Data was collected through quantitative surveys of 400 households and 400 individuals in 2007 and 304 households and 288 individuals in 2010. Both waves of quantitative data collection were complemented by qualitative data collection through focus group discussions, key informant interviews and case histories using interview schedules. The study was guided by a conceptual framework that was a modification of Duncombe's (2007) extension of the livelihood approach. The study reports mixed results with respect to the relationship between ICT access and usage and the different dimensions of poverty. The chapter ends with some conclusions and recommendations for policy interventions.

As part of the major study on assessing the impact of Information and Communication Technologies (ICT) on development and economic transformation, Peter Ulanga has explored the role of ICTs in development and economic transformation especially in rural areas of Tanzania, by looking at the users and the sustainability of the telecentres. The research findings are discussed in Chapter 8. The study found that ICTs are used primarily as means of sharing information in the rural societies and they contribute to improvement of social economic development. Significant level of ICT use for economic activities is observed among people from different income brackets and different education profiles in the rural areas. However, a greater impact of ICTs is experienced on the users in rural areas who are more educated (secondary education leavers) compared to the less educated (primary education leavers).

Chapter 9 by John O. Adeoti, Foluso M. Adeyinka & Moses O. Ubaru examines the impact of the Nigerian policy on information and communication technology (ICT) on printing and publishing industry. The data are obtained from a survey of the industry and three case studies carried out in 2009. The findings show that 68% of the respondents are aware of the existence of the ICT policy, but some of them are ignorant of its contents. The ICT policy had impacted at least 40% of the respondents in a positive manner, such as improvement in market for products, introduction of new or improved products, adoption of new equipment/machines, made investment in ICTs less expensive, and stimulation of more investment in ICTs. This not withstanding, the case study illustrations suggest that the ICT policy impact is still not deep, and firm expectations in the printing and publishing industry are still far from being met. It would therefore be necessary to further strengthen the implementation of the ICT policy in the Nigerian printing and publishing industry.

The government has recognised the importance of ICT in development, but all efforts are geared towards expanding ICT use and nothing to encourage local ICT development/production, especially application software as evidenced in national ICT strategy. There is a niche for local developers to satisfy local needs/export as the application software market is still largely unexplored. Using a modified Porter's Five Forces Model for Industry Analysis, in Chapter 10 Sunday A. Khan and Arsène Gidéon Nkama examine the application software market and assesses the difficulties faced by local application software developers. The banking sector is dominated by foreign banks using imported software, consequently beyond domestic competitive forces. In the microfinance sector on the contrary, local software dominates the market, though competition is severely hampered by membership of business associations/networks that influence choice of application software. There exist some ambitious and well-structured local developers facing difficulties in software development, marketing, government patronage and piracy. In order to increase impact of local application software development on economy, government should fight piracy, patronise local firms, create a service to promote the local application software industry, and improve the ICT regulatory/institutional environment.

Development of new technologies was not always accompanied by an improvement of labour productivity. Therefore, Latif Dramani and Oumy Laye Under analyse the impact of the investment in ICT on the economic growth and employment in Senegal in Chapter 11. The methodology employed by the authors takes as a starting point the model of social account matrix multiplier, Pyatt and Round (1979), Defourny and Thorbecke (1984), and the multipliers derivation initiated by Thorbecke et al. (1996), aiming to measuring the possible externalities and impacts that investments on ICT generate on poverty reduction and jobs creation. The results highlight a significant effect of the ICT investment on growth, employment and poverty alleviation in Senegal.

Chapter 12 by Vijay K. Bhasin, Simon K. Harvey & Eme Umoeka Fiawoyife shows that the ICT investment in cellular phones by the commercial taxi and fishing-boat firms do not have any effect on their productivity, whereas it does have some effect on their total cost-wage ratio. The study shows that there is plenty of scope to raise the technical efficiency and to reduce the allocative efficiency of commercial taxi and fishing boat firms. The study also establishes that fishing industry is less efficient than the transport industry. The technical and allocative efficiencies of commercial taxi firms can be changed by encouraging younger people to start the commercial taxi business, whereas the technical efficiency of the fishing-boat firms could be increased by encouraging educated people to start the fishing business.

Most of Africa's peoples live in rural areas where agriculture is the main economic activity, dominated by small enterprises, traditional technologies and low productivity. Transforming agriculture with modern technologies, including information and communications technologies (ICT), is the required strategy for improving rural and national wellbeing in African countries. This perspective motivated the study by Mutawakilu A. Tiamiyu and Abiodun S. Bankole & Rosemary O. Agbonlahor (in Chapter 13), which investigated the patterns and impact of use of ICT in cassava value chains in South Western Nigeria. The study performed regression analyses on primary data collected through in-depth interviews of 139 cassava growers, processors and marketers. ICT ownership bore a direct relationship with output values, but neither ICT investment nor use could predict output values or output-input ratios. Recommended are strategies to encourage young and educated youth to venture into value chain activities, and to promote the expansion of cassava growing, processing and marketing thereby catalyzing more diverse beneficial uses of ICT in the value chain.

In Chapter 14 Silvia Masiero conceptualizes the academic field of Information and Communication Technology for Development (ICT4D) as constructed by the building blocks of information systems and development management. Given this double genesis of the field, the concept of accountability in ICT4D should embody the complementary notions of accountability in information systems, which is internal and upward-flowing within organizations, and accountability in development management, which is external and downward-flowing towards development beneficiaries. The author used the fieldwork in Kerala, on a telecentre project oriented to human development, to illustrate the functioning of both accountabilities in practice, and the mechanisms through which they mutually interact in a relationship of cause and effect. The author finally concludes that accountability to beneficiaries may depend on internal cooperation and coherence of development organizations, and therefore, ICT4D projects should be studied under both dimensions of accountability, which complement each other in designing effective interventions.

Finally, in Chapter 15 Fatuma Simba, Lena Trojer & Zaipuna O. examine the availability of supporting environments for the provision of broadband rural connectivity. The surveyed environments include initiatives for ICT infrastructure development and institutional frameworks such as policies, strategies, legislations and regulations. The aim of the survey is to investigate the readiness of Tanzania to deploy Public Private Partnership model for rural connectivity implementation. Furthermore, the study proposed implementation strategies in order to adapt the Public Private Partnership model for sustainable broadband rural connectivity in Tanzania.

In conclusion: all contributions highlight that government policies and private sector participation are critical in stimulating, monitoring and evaluating progress in the ICT industry and its role in both economic growth and meeting the broader social objectives. Though the countries across Africa have different levels to their challenges, the nature of the challenges is similar.

Chapter 2
ICTs and Economic Transformation in Africa: a Synthesis

Olu Ajakaiye & Samuel Wangwe

1. Introduction

The process of technological innovation from the late 20th century and its acceleration in the 21st century has been dominated by new technologies and in particular information and communication technology (ICT). Information and communications technology (ICT) refers to technologies that pertain to the new science of collecting, storing, processing and transmitting information whereby information, computing, and telecommunications are converging. The ICT sector is a combination of manufacturing and services industries that capture, transmit and display data and information electronically that is breaking the traditional dichotomy between manufacturing and services making it possible for activities producing and distributing ICT products to pervade all sectors of the economy (OECD, 2003).

ICT can be thought of as an integrated system that incorporates the technology and infrastructure required to store, manipulate, deliver and transmit information, the legal and economic institutions required to regulate ICT access and usage, and the social and inter-personal structures which allow information to be shared, facilitate access to the ICT infrastructure, and through which innovation takes place. Three broad sub-systems make up the ICT system, each of which comprises further sub-systems: technology and infrastructure (comprising communication and data processing); institutions (comprising legal and economic institutions) and social structures and processes (these influence ICT usages and the inter-personal links that facilitate the transfer of information). This interpretation of ICTs paves the way for understanding the multi-dimensionality of ICTs and their applicability across various sectors.

The world is experiencing a new industrial and technological revolution which is bringing about a significant, fast and extensive transformation of society and economic activity resulting in the rapid transformation of the processes of production and the transmission of goods and services produced. The ICT systems are pervading virtually all forms of human endeavour: work, education, leisure, communication, production, distribution and marketing as well as the content of information networks. They play a major role in all aspects of national life: in politics, in economic life, and in social and cultural development. It is transforming the lives of people, the way they do business, access information and services, and communicate with each other. It provides benefits relating to governance and human rights by supporting freedom of expression and the right to information. ICT is a pervasive input to almost all human activities and breaks barriers to human development by providing content to knowledge, breaking barriers to participation and to economic opportunity.

ICTs have diffused into all branches of the economy and their impact has an impact on virtually all aspects of productive activity in the economy. The nature and extent of ICT adoption has been explored quite extensively in recent research, though in Africa understanding of the impact of ICTs on economic transformation is rather limited.

The African Economics Research Consortium (AERC) project on ICT and Economic Development and Transformation has examined the implication of various kinds of ICTs on economic transformation in Africa based on case studies in 10 African countries.

More specifically the project addressed two main questions: First, it identified the status of ICT access and adoption in production and growth generating activities that could lead to economic transformation in Africa. Second, it explored options and possibilities of using ICT to enhance economic development and transformation in Africa.

2. ICT Evolution

The world has undergone an evolution in terms of structural change and technological transformation. However, Africa is still lagging behind in this modern world whose development is largely determined and led by technological advances. Increasingly, ICT plays an important role in determining levels of development and the quality and accessibility of services in our modern societies. It is in this context that governments and world leaders made a strong commitment towards building a people-centred, inclusive and development-oriented information society for all, where everyone can access, utilize and share information and knowledge.

A major concern is the risk of the digital divide between those who have access to digital technology and are benefiting from the digital age, and those who do not. This great divide between rich and poor countries and between rich and poor sections of societies within countries is developing and reinforcing the long observed inequalities in economic wealth and social conditions.

It is argued that the historical perspective of the technological revolution provides Africa (and other developing countries concerned) with a basis for leapfrogging and to use ICTs to all the economic activities such as industry, agriculture, social services and government administration. The experience of the now developed countries can be useful in terms of the core institutional structures that are needed to facilitate and regulate the development of ICT and the options for transforming the African economies.

3. Status of ICT and its Diffusion in the African Economy

The African ICT sector is undergoing tremendous advances, which have changed people's lives and the way they participate in the development activities using various information and communication devices such as mobile cellular phones, radios, faxes, televisions and computers.

Where incomes are low, infrastructure is undeveloped and population density is low, challenges to accessing ICT are great though not insurmountable. Alternatives to individual access are being developed in terms of facilities shared by communities such as telecentres, cyber cafés and community information centres. Strategies to draw people to use these facilities are essential for developing African countries.

Hundreds of new local and community radio stations are now operating and satellite TV is now also widely available on the continent. Communication is essential to rural development with importance now attached to grassroots participation and sustainable development. Rural radio is the most inexpensive. It is the popular and socio-culturally appropriate means of communication, being also the most accessible to most of the population in Africa. Radio is still by far the most dominant mass medium in

Africa with ownership of radio sets being much higher than that of any other electronic device. The community radio also provides a unique space for women, where they can express themselves in the language which their particular community can understand. Due to the convergence of technologies some community radios are broadcasting on the Internet and helping women to reach even broader audiences.

Satellite-based broadcasting has seen major activity on the continent in the last few years. For example, many countries in Africa are now connected to the South African company M-Net, which deploys the digital direct-to-home subscriber satellite service called DSTV. This type of service provides access to several video channels and audio programmes to the whole of Africa at an affordable cost for many middle class people in the continent.

The use of ICTs has grown relatively rapidly in most urban areas in Africa with the Internet spreading to every major city. However, the differences between the development levels of Africa and the rest of the world are much wider in this area. More mobile cell phones are now deployed on the continent than the number of fixed lines laid in the last century. The number of mobile phone subscribers in Africa has increased rapidly in recent years.

Computers are still by far the most important gateway to the Internet even though the Internet is increasingly being accessed through a variety of devices such as telephones in recent years. However, computers continue to be indispensable for the development of the information economy and in particular for the application of ICT in e-business processes. Computer penetration rates are lowest for Africa. Although broadband access to the Internet has become a regular feature in the enterprises of developed countries, the number of broadband subscribers in most African countries is extremely small, and penetration rates are less than 1%.

In a brief analysis of African government websites, more than three-quarters of all countries in Africa have a web presence. However, challenges of content, updates and widespread use of the information in the web continue to be of major concern. Most sites have an external focus as demonstrated by the type of information provided (tourism, business/investment opportunities, and general introduction to the geography and culture of the country). Most sites are in English rather than in African languages. Many sites do not yet have interactive formats with most sites functioning as vehicles of disseminating information rather than two-way information flow. Only about a fifth of all sites have any significant two-way information flow or interactivity features.

The realization of the potentials of ICT in Africa in most cases is hampered by challenges of infrastructure (physical and regulatory), lack of access to technology in rural or remote areas and to the poor and the underprivileged (generally women and minorities); prevailing illiteracy, both computer-based and otherwise, and lack of content in local languages further aggravate the difficulties.

Africa is taking up the challenge of reforming the telecommunications sector in an effort to make it capable of supporting broad-based national economic and social development. As a result of telecommunication sector reform, many parts of the continent are making improvements in the telecommunications network.

The developments in the telecommunication sector have shown that leapfrogging is possible in Africa as evidenced by the prevalence of wireless over fixed line communication devices. The mobile technology brings communications to whole communities that previously had little or no access to fixed line telephones.

The liberalization of the ICT sector has resulted in an increasing number of players in the sector such as investors in telecommunications, ISPs, and cellular phone provider operators, a situation which calls for the capacity to regulate the multiple players. Thus the number of ICT regulatory authorities in African countries has increased dramatically in recent years to referee the market place. However, their capacity is a major challenge.

Africa's competitiveness is heavily dependent on the skills of its people. The effective use of ICT and its spread is limited by the level of requisite human skills. The future of ICT in the continent depends upon its ability to develop the skills to cope with the changing ICT environment.

ICT policies in Africa have addressed a wide range of issues such as market structure, access network, licensing, universal competition policy, and essential ICT services. It is pointed out that the ICT policy and regulatory process in Africa has continued to pursue the international reform of the communication sector that began in the early 1990s with relatively less emphasis on policy goals of delivering affordable access to the population and other reform elements such as competition and effective regulation. Little attention was paid to institutional capabilities, resources, markets and governance and policy coordination issues at national levels. It has been observed that some countries have made encouraging progress in creating comprehensive national ICT policies, strategies and plans to deliver broader socio-economic outcomes through the appropriation of ICTs.

There has been a significant divergence between sector reform agendas and efforts to develop national e-strategies in Africa. Sector reform efforts failed to recognize the implication of integrated ICT policies, while national ICT strategies fell short of capitalizing on national reform efforts towards competitive environment for affordable access. These failures could have serious implications for Africa's integration into the global information economy. The political economy of the ICT sector reform was shaped both by external factors, policy choices, domestic power relationships and vested interests of different groups.

In spite of a large number of countries that have formulated integrated policies and the rhetoric of commitment at national and international levels, the vast majority of African countries have yet to optimize the potential of ICTs due to weaker markets, financial constraints, limited human resources and ineffective institutions. ICT policies and plans that were developed largely by ICT experts with little involvement of other stakeholders have experienced shortfalls at the implementation level. The main challenges have included governance and leadership, investment climate and incentive structures for private investment.

4. ICT Impact on Economic Development and Transformation

ICT can contribute to income generation and poverty reduction, and can facilitate economic transformation in Africa. The use of ICT to enhance economic development and transformation in Africa entails addressing the impact of ICT on changing the structure of the African economy (composition of agriculture, industry and services), expanding economic and social development opportunities, facilitating diversification, exploring options for building competitive advantages, facilitating efficient functioning and responsiveness of institutions (including markets) with a view to creating vibrant markets and institutions.

Technological leapfrogging offers an opportunity for developing countries to catch up with modern ICT resources. They provide more cost effective and user friendly options. The impact of ICTs on development and economic transformation can be mediated through several channels. These include its impact on:

- The ways and the speed of acquiring information and knowledge (impact on education and learning) across societies and sectors.
- Production and the way the production process is organized whereby the technology impacts on the organization of the work place much in the same way that electricity brought about different modes of production. ICTs are affecting on the way businesses are organized.
- Productivity and its consequent impact on wealth creation by transforming traditional production systems as well as creating new production regimes.

- Networking whereby linking people, places and events all over the globe occurs revolutionalizing the way various actors interact and relate to other actors in the country, region or globally. This impact includes the way business and research are carried out.

5. The Future of ICT and Implications on Economic Transformation

It should be recognized that there is a time dimension to IT investment because of the learning effect of technological investment within which firms master techniques and by which "network effects" begin to be felt. Policy makers should therefore plan for lags in investment. For instance, adoption of advanced manufacturing practices may require significant changes in work organization that may sometimes be disruptive while making positive impact on productivity.

The impact of ICTs on development and economic transformation may be determined at the aggregate level or at a disaggregated level. Analysis at the aggregate level would try to establish, at the macro level, the impact of ICT on economic growth or on economic transformation. The rationale behind the aggregate approach is that the dynamism of ICT is expected to come from several sources including the decline in the prices of information processing, convergence in communication and computing technologies, and the rapid growth in network computing. The studies which have adopted the aggregate approach have largely been done on the developed economies. They have found that the OECD countries that improved performance in the 1990s were generally able to draw more people into employment, increase investment and improve factor productivity (Pilat et al., 2002). The paper by Dhramani and Laye in this special issue is one of the very few studies which have addressed the aggregate impacts on the African economy. Their study on macroeconomic impacts of ICT on the economy of Senegal showed a significant effect of the ICT investment on growth, employment and poverty alleviation. The analysis of the impact of ICT investment on the fundamentals of the economy has shown dynamism and scope for the rise in incomes, employment and poverty reduction

The impact analysis therefore has been done at a more aggregated level. Two types of disaggregation are adopted. The first is disaggregation in terms of types of ICT. The main categories are ICT producing activities and ICT using activities. The second type is by category of users (individuals, households, businesses and government).

Basic Social Services

Provision of basic social services like health and education is being facilitated by ICT development in Africa.

Education

Across a range of educational applications, ICT is being harnessed to improve the efficiency, accessibility and quality of the learning process by enabling increased access to knowledge and adoption of more collaborative and interactive learning techniques[1]. ICT enables distance learning and effective adult education. Multimedia technologies such as radio, projectors, public address systems, CD-ROMs and television are making their way into educational institutions.

Distance learning is reducing barriers of affordability and geography, especially in tertiary education. Through distance learning, use of educational software and IT-related professional training programmes, ICTs has facilitated wider access to appropriate culturally educational and job training, thus producing a work force with better skills.[2] Many universities offer non-credit, remedial and specialized certification programmes via satellite-based educational networks. For example, the University of Namibia offers a distance learning bachelor degree programme in nursing in which women constitute between 90–95% of the students.[3]

Health

Health systems such as the HealthNet have had a significant impact in most countries in Africa. The HealthNet system of local telecommunications sites are used to provide low cost access to health care information in developing countries via the Internet. Users, mainly physicians and medical workers, can connect to the network through local telephone nodes to access services such as physician collaborations, medical databases, consultation and referral scheduling, epidemic alerts, medical libraries, e-mail and shared research reporting databases. Many physicians in developing countries rely on HealthNet as their sole source of information on the treatment of AIDS and tropical diseases, essential drugs, paediatrics and public health promotion. For example: surgeons from Mozambique, Tanzania and Uganda can consult on patient treatments and learn new re-constructive surgery techniques.

Health and education departments in many countries are beginning to electronically transmit operational MIS statistics such as disease occurrences and pupil registrations. In South Africa, the results of blood tests are being transmitted to remote clinics that are off the telecom grid via mobile telephone text messages.

Some of the most promising and clearly demonstrated applications for ICT in development are in the improvement of health care delivery. In many developing countries and communities, it is used to facilitate remote consultation, diagnosis and treatment. In Gambia, for example, nurses in remote villages use digital cameras to download images of symptoms onto a PC and transfer them to nearby towns for examination by doctors. The same model is being applied to facilitate collaboration among physicians. When an expert opinion is required, doctors in rural towns send the images captured by the nurses to specialists in the United Kingdom for advice. On the issue of community sensitization and education, radio, television and Internet are used in African countries to promote HIV/AIDS education and prevention.

The principle of ICT-facilitated collaboration also extends to medical research. This is illustrated in West Africa, where malaria researchers use a network of satellites and ground stations to submit data for clinical trials conducted at tropical disease research facilities in London and Geneva.

In the health sector, GoR established specific programmes related to health sector. These include[5] an open-source medical records system that tracks patient-level data, TRACPlus and TRACnet which monthly monitor infectious diseases, a network to facilitate inter-hospital connectivity and ensure easy accessibility to health services and an unlimited exchange of data between hospitals.

In the rural area, Business Communication Services (BCS-Tuvugane) intervenes in serving the rural and most remote areas with no mobile phones. Tuvugane phones enjoy lower tariff structure than ordinary mobile phone calling costs. The second service provided in rural area is village phone or tel'imbere. In partnership with Rwanda's leading microfinance institutions, Village Phone Rwanda creates opportunities for poor rural individuals to become "Village Phone Operators" or tel'imbere operators. These tel'imbere businesses can be established in areas where electricity is unavailable and in areas where the MTN and Rwanda tel networks can only be accessed with a booster antenna.

Agriculture and Rural Development

Lack of timely information is known to be the largest constraint to small-scale agricultural production and natural resource exploitation—a sector that provides livelihood for 70–80% of Africa's population. However, thus far the potential for ICTs to have an impact on this sector has not yet received much attention.

The emerging new technologies are transforming the way production is organized in Africa's agricultural and rural development. ICT offers means of exchanging and mass distributing advances in farming techniques, pest control and agri-exports. Telecentres are facilitating distance learning and enabling farmers in the rural areas to use ICT to access knowledge, share information and acquire farming skills to enhance their crop production, thus creating a culture of information and experience-sharing within African communities.

To address the challenges of land degradation, remote sensing data is being used to produce maps to show areas with potential soil erosion hazards, thus facilitating decision makers to take the necessary interventions to ameliorate the situation. The provision of public information and easy access is one of the most direct routes to broad participation in environmental management and better environmental decisions.

In Kenya, mobile phones are helping farmers obtain a fair price for their produce in the country. Vodafone's associate company in Kenya, Safaricom, has introduced a text messaging service which provides quick and easy access to updates on agricultural markets. Users simply text the name of the commodity they are interested in and they receive an instant reply with an update on its price that morning at each market. This enables them to assess the best place and time to buy or sell.

In the case of Rwanda, the Ministry of Agriculture set up a network bringing together all stakeholders involved in agriculture sector using an exchange platform, another network - Agricultural Management Information System (AMIS) to enhance interaction among stakeholders has been set up in order to link farmers to markets.[6] E-soko project seeks to empower farmers to enable them make more informed market pricing decisions and ultimately more successful farming.

A study of Tanzania by Peter Ulanga (see Chapter 8 of this book) has found that the use of ICTs in rural areas in the telecentres is influenced by the level of education especially for internet and computers. It is interesting to note however that there was no trend when the use was compared to income levels, suggesting that ICTs are considered essential and not products of excess liquidity or luxury at the individual level. However, considering the utilisation of the computers is still on the low side, similar to earlier studies (Souter et al., 2005), it was found that information sharing has been the most significant use of ICTs. There is a significant use of ICTs directly for economic activities in the rural society which depends on the level of education of the users of ICTs. It has been shown that telecentres are marginally sustainable and not providing types of services that will improve utilisation. For quite sometime to come radio and television are expected to continue to be the most prevalent means to reach out population for information purposes. However, it is possible to change this scenario if deliberate policy is directed to reach out the general population by locating telecentres in public places like community centres at ward level or lower levels or at schools in the rural areas. The economic linkages have yet to be realigned to increase the value of ICTs in economic activities in order to increase economic value of the ICTs in rural areas. This is a potential that could be harnessed more effectively.

The case study on value addition in cassava industry in Nigeria, presented by by Tiamiyu, Bankole and Agbonlahor (see Chapter 13 of this book), found that cassava growers are more inclined to seek credit and undertake risky investments than the processors and marketers. South-western Nigeria is among the most literate and economically developed regions of the country. The sample probably contained a higher proportion of wealthier members than one would find among Nigeria's population. The important predictors of ICT ownership, investment and use among growers, processors and marketers include age of firm, and the firm owner's gender, age, education, knowledge, credit seeking and extent of participation in trade and credit associations.

ICT use is likely to help growers both in sourcing for improved inputs and growing methods, and favourable prices for inputs and output. For processors however, ICT use is more likely to help in sourcing for favourable prices for cassava produce and output of cassava products, and less likely to help in the produce processing aspects of their business, as the technology is currently mostly basic and stable. Marketing, being a service activity, provides the widest scope for the profitable use of ICT by providing better access to market information than either growers or processors, and marketers often use this advantage for arbitrage or speculative purposes. However, ICT use among the cassava value chain participants was limited to basic and low-end ICT, primarily radio, TV and cell phones. Women-dominated processors were the least educated compared to the growers and marketers, and were least inclined to explore markets and engage in frequent market transactions. What is needed is to have programmes to promote women-controlled cassava processing associations or societies that engage in thrift, credit and collaborative investment, processing and marketing activities. This study on cassava processing in Nigeria found significant relationships between ICT ownership and output values among processors and marketers. The study also found significant correlations between the ICT-related variables and the prices obtained by growers and marketers for cassava produce and garri product respectively.

Industry

Oyelaran-Oyeyinka and Lal (2004) investigated the impact of ICTs on industry in Africa and found that the intensity of ICT tools adoption was not affected by factors such as profitability, size of operation, age of firm, and per capita investment at the industry level. However, there are significant variations in the conduct and performance of firms that used the lowest levels of e-business tools from the more advanced users of new technologies within an industry (Oyelaran-Oyeyinka and Lal, 2004a). It is suggested that supply side factors significantly affect the adoption of new technologies in SMEs. The most important factors are availability of physical infrastructure, technological infrastructure as represented by availability of Internet connectivity and speed of communication and human knowledge and skills represented by availability of a computer literate work force.

The study by Adeoti, Adeyinka and Ubaru (see Chapter 9 of this book) addressed the impact of the national ICT policy on printing and publishing industry in Nigeria and found that 68% of the respondents are aware of the existence of the ICT policy, but some of them are ignorant of its contents. The ICT policy had impacted at least 40% of the respondents in a positive manner, such as improvement in market for products, introduction of new or improved products, adoption of new equipment/machines, made investment in ICTs less expensive, and stimulation of more investment in ICTs. However, it is apparent that there is a need to further strengthening the implementation of the ICT policy with respect to the printing and publishing industry.

The study by Bhasin, Harvey and Fiawoyife (see Chapter 12 of this book) has addressed the boat fishing industry and the taxi industry in Ghana and found that ICT investment in cellular phones is widely used by the commercial taxi and fishing-boat firms but it does not yet show any significant effect on productivity or on the total cost-wage ratio. The study suggests that there is plenty of scope to raise the technical efficiency and to reduce the allocative efficiency of commercial taxi and fishing boat firms if younger and better educated people could be engaged in the industries.

Khan and Nkama (see Chapter 10 of this book) addressed the challenges of development of software industry in Cameroun and found that government policy is directed more on ICT use than ICT production. Although there is a niche for local software developers to satisfy local needs/exports, the application software market is constrained by local factors. For example, its development in the banking sector is inhibited by the dominance of foreign banks using imported software. In the microfinance sector which is dominantly locally owned the local software dominates the market, though competition is severely hampered by membership of business associations/networks that influence choice of application software. Although, there are some ambitious and well-structured local developers, they are facing difficulties in software development, marketing, government patronage and piracy. It is found that the government has not done enough to fight piracy, patronise local firms and improve the ICT regulatory/institutional environment that is necessary for the development of the industry.

Tourism

Globalization and new ICT are radically transforming the tourist industry. The growing use of the Internet is enabling Africa to tap the potential of the tourism industry in terms of tourism promotion, marketing and sales. ICT is providing an opportunity for Africa to have effective means for marketing and selling their own tourism services online, especially through the Internet. This is increasing the capacity of these countries to respond fast to the needs of today's travellers for quick access to information, customized software products and secure, online payment facilities for travellers. ICT has enhanced the capacity to present information on their tourism offer, and match it with the worldwide demand. It includes locally produced information about the tourism offer of the countries, and value-added online services such as hotel reservation. With e-tourism, official guides of the country's hotels are being provided with information on hotel rooms with facilities for online reservation.

Trade

The web-based systems are becoming a fast-moving sector in the world, as they are increasingly being used in commerce. Traders are using ICTs such as telephone, e-mail and websites at the workplace to maintain business relations with partners outside their respective countries. The traders have acknowledged that outlets for their goods and services have increased. The information such as markets, market regulations, prices and potential suppliers and buyers may be posted on the web for easy accessibility. There is an increase in the use of other technologies such as electronic shopping facilities and credit cards to enable traders to gain competitiveness and unprecedented opportunities offered by the digital revolution for economic development.

Governance

ICT is supporting and facilitating governance as the process through which institutions, businesses and citizen groups articulate their interests, exercise their rights and obligations and mediate their differences. One important role of ICT is the enhancement of e-governance for rural development in Africa. It is facilitating provision of basic social services, improvements in public administration and enhancing democratization. There is a notable increase in official general government websites providing information that was not available before. As greater numbers of public officials are now gaining low-cost access to the web, the vast information resources available via the Internet are becoming increasingly important tools in ensuring informed decision making.

The experience of Rwanda illustrates how ICT can help transform the public service. As a part of e-Government project, the Government of Rwanda has started to install Internet Kiosks in different districts. The project is aimed to encourage the use of ICT in public service delivery. The case of Rwanda has shown that ICT has transformed the manner in which the Immigration Department is delivering services in the case of processing applications for passports. The process has been shortened from 20-30 days to 4-6 days. Application forms are downloaded on the website of Directorate of Immigration and Emigration, and be submitted online. After a few days, the person requesting a passport gets an e- sms on the mobile phone informing him or her that the passport is ready for collection. The results from the survey show that the persons are satisfied with the new process of Rwandan passports delivery.

Thus, the modernization of ICT systems has the potential to facilitate the transformation process if it is integrated into the process of organizational and societal change in Africa. ICT can facilitate growth, basic service delivery, enhance monitoring and evaluation, building awareness of critical issues (such as Sanitation, Health, Education), private sector development (development of content providing and SME development) and governance (public procurement, public service delivery, providing a knowledge base for policy monitoring and evaluation, promoting open and participatory processes) and transforming agriculture through information systems; price information systems; increased market access; increased production efficiency; and empowerment in conducting economic activities. In order to optimize the use of ICT in the transformation process greater effort is needed to collect information and develop conceptual understanding about the relationship between ICT and socioeconomic transformation. The role of ICT in economic transformation has to be considered in the context of the national and regional development agenda, a basis for providing a working fit between the global system and national and regional capabilities and strengths.

Promoting transformation through value-added services poses a major challenge for Africa if the continent is to carve for itself a niche in this part of the "value-chain". Production and use of ICT stand on the pillars of information infrastructure, human capital, and an innovative system. An appropriate policy regime for promoting production, trade and investment is essential for facilitating the impact of ICT on economic transformation in Africa. Various actors such as SMEs need institutional support to facilitate their movement up the value chain and participate more effectively in the era of globalization and rapid technological advances.

Development of capabilities to produce and use the appropriate information is a major challenge. Adoption of content to reflect local conditions and the use of language that is widely understood in African countries is a major challenge. The bulk of the appropriate information will need to be locally generated and managed, that is, gathered, analysed and synthesized, maintained and made available in a form beneficial to the interested parties. The categories of information that are relevant in the African context include market information (costs, prices and suppliers of products and services); research, development and experiential information; regulatory and normative information (information on standards, norms and good-practice); and performance information (information on actual quality of products and services or actual performance levels of industries and institutions, etc.).

Policy initiatives will need to be taken to grapple with the challenges of spreading the use of ICTs, especially to the rural areas in Africa in order to address the digital divide. The governments should take deliberate action to address the wide urban-rural digital divide and the large-small enterprise divide which is more accentuated by the complexity of the technology in use. The gap is wider in the area of internet use and broadband use. Effective use of ICT should be promoted among enterprises especially the small enterprises so that they can use computers not only for e-mails but also for raising productivity in production, marketing, accounting and banking transactions. In this regard, priority should be accorded to leveraging the internet more effectively by encouraging and facilitating enterprises and other users to rethink some business and productive processes around new ICT tools and invest in ICT skills development. Small enterprise managers may be taught how to use relevant software packages to perform accounting, financing and strategic planning of their projects and such training could be offered at subsidized rates. Action should be taken to reduce the digital divide between urban and rural areas by raising the level of literacy of rural dwellers. For instance, training farmers to use software which suites most of their farming activities would enable the farmers to record, retrieve and save farm records on their office computers. Farmers can be taught to uses ICT facilities such as computer, fax, mobile phones and internet so that they may take advantage of these facilities and perform more effectively. A livestock farmer may report animal deficiencies to relevant authorities through internet or mobile phone messages and get immediate attention.

The level of awareness of ICT needs to be promoted especially in the countries and regions by continuously conducting awareness and sensitization campaigns for making people understand the critical role of the sector to the society and the economy in general. A high level of awareness on ICT will enable the public to know opportunities of using ICT for improving their standard of living, for bridging the digital divide, and for the transformation of the African economies. Promoting awareness on ICT will help to break the vicious circle of low demand, low investment in ICT, high prices of ICT services and low demand. ICT awareness will stimulate demand for ICT services which will stimulate further investments in ICT which will facilitate a faster fall in the prices for ICT services which will encourage further demand for ICT services. Awareness and capacity building remains one of the major challenges to ICT policy formulation and implementation in Africa.

Human resource development is an important dimension of the existing knowledge gap. In today's world, knowledge gaps may be filled through international educational and technological collaborations and strategic industrial alliances. Given such opportunities, what is important for African countries therefore is to set appropriate developmental goals, realistic and achievable over a given timescale and within available resources, to identify the gaps in knowledge, and to concentrate efforts to remedy the situation, that is, to acquire, assimilate, adapt or refine the missing, or the insufficiently developed, areas of knowledge. A major challenge is that of establishing the information needs and skills of the majority of users if ICTs are to effectively enhance market integration, especially in the rural areas in Africa. The status of literacy of ICT and output of ICT trained/skilled personnel needs to be raised to cope with current and future requirements of the sector. The available training in ICT may require realigning of the educational and vocational training to meet the needs of the fast evolving labour market. Action should be taken to raise the level of ICT literacy as well as stepping up information literacy in the country. This will require that ICTs are fully integrated in the education and training system at all levels including in curriculum development, teacher training and institutional administration. There is need to increase computer training to primary and secondary thus creating a good background for them and raise awareness which is needed for more effective utilization of the computers. Investment in human resource development for ICT should be undertaken by the public as well as the private sector. Avenues for public-private partnership should be explored along with the provision of incentives to encourage the private sector to invest in human resource development for ICT.

A major concern relates to the implications of ICT for women and other disadvantages groups in society. Although the potential for ICT to serve as an empowering force for these groups is real, the currently low participation of these groups at the ICT kiosks suggests the risk of further marginalization unless deliberate initiatives are taken to address the challenge. Policy for promoting ICTs should ensure that ICTs are highly accessible to many people in Africa making it possible to step up the innovativeness in content of educational programs that are relevant to the development challenges. Priority should be accorded to investment in public ICT facilities such as public internet facilities in telecentres in communities, schools, internet cafes and public libraries. This will call for capacity building programmes on ICT to be carried out on how to use internet and other ICT components more effectively. Initially action should be taken to recruit computer officers who will provide the lead in these public ICT facilities so that even those who are currently illiterate in ICT can be assisted to make effective use of the facilities. For instance, a computer official may surf the internet and find relevant information to be communicated to the needy in the respective communities.

Infrastructure is central in achieving the goal of the digital inclusion, enabling universal, sustainable, ubiquitous and affordable access to ICTs by all, and to provide sustainable connectivity and access to rural areas. Sound and sustainable national ICT Broadband Infrastructure Backbone (fibre-optic) network is essential for the sector to deliver the services. Infrastructural support is necessary for the adoption of advanced e-business technologies and in order to reap the benefits of ICTs and develop the capabilities to contribute to economic development. Policies and programmes aimed at providing required infrastructure need to be initiated in African countries in order to make various economic activities more competitive in the domestic and international markets. One such policy is the provision of collective services.

One important area of future concern is the prospect of improving official statistics on ICT, and making the available data more comparable internationally. In order to be able to assess the extent, constraints, and potentials of using ICT in improving livelihoods of the poor and in supporting growth in general, it is recommended that a sufficient number of sector performance indicators be identified that are suitable for monitoring the sector performance. The indicators to be developed will help to monitor developments in infrastructure and in the use of ICT in terms of access and utilization. By trending these indicators, it would then be possible to make an assessment of the socio-economic impact of ICTs on the livelihoods of the poor as well as the sector impact on the transformation process. National and regional surveys should incorporate ICT data collection that can generate the desired indicators. In developing such performance indicators, it is advisable to follow internationally comparable indicators which are being developed in the Partnership on Measuring ICT for Development.

There is room for increasing regional cooperation and integration among African countries to fully harness the potential of the ICT for economic development. There are options for learning from experiences of good practices from other countries but these have been limited and are yet to be effectively harnessed.

The governments should adopt policies and regulations to facilitate broadband rollout range from taxation and fiscal incentives to market liberalization and may also include universal access and market stimulation.

In summary, we recommend following policy initiatives:

- Operators are encouraged to share state-of-the-art backbone infrastructure to avoid duplicative and fragmented low bandwidth networks.
- Governments should promote the use of ICT in business by creating an enabling environment for e-commerce and e-business and promoting literacy among customers and suppliers.
- Policy should promote the skills of the workforce and their level of literacy.
- Policy should promote local content which is relevant for locale specific development.
- Take action to speed up the allocation of wireless spectrum with a view to overcoming the broadband access gap.
- Promoting competition among operators and reducing barriers to entry.
- Taking action to stimulate the market to grow.

The governments will need to review the legal and regulatory framework on a continuous basis to more effectively accommodate electronic transactions and improve consumer and business confidence in their use. Such steps could include the legality of the use of electronic documents, e-signatures, privacy, intellectual property rights, cybercrime and building trust. These steps can widen market access for SMEs nationally, regionally and internationally

Notes

1 Available at: http://www.opt-init.org/framework/pages/2.2.2.html.
2 Available at: http://www.uneca.org/adf99/governance.htm#10#10
3 Available at: http://www.uneca.org/adf99/governance.htm#11#11
4 Available at: www.nepad.gov.rw
5 Hamish Frasier et al., e-Health Rwanda Case Study, July 2008, p.2
6 Available at: MINAGRI, http://amis.minagri.gov.rw/content/information-gateway-agricultural-and-livestock-sector-rwanda

References

Oyelaran-Oyeyinka, B. and Lal, K. (2004), 'Sectoral Pattern of E-business Adoption in Developing Countries,' *Discussion Papers 07*, United Nations University, Institute for New Technologies, Maastricht, Netherlands.

Oyelaran-Oyeyinka, B. and Lal, K. (2004a), 'Determinants of E-business Adoption: Evidence from Firms in India, Nigeria, Uganda,' *Discussion Papers 14*, United Nations University, Institute for New Technologies.

Pilat, D., Lee, F and van Ark., B. (2002), 'Production and use of ICT: A Sectoral perspective on productivity growth in the OECD area', *OECD Economic Studies No. 35*, Paris: OECD.

OECD (2003), *ICT and Economic Growth: Evidence from OECD countries: Industries and Firms*, Paris: OECD.

Souter, D., Garforth, C., Jain, R., Mascarenhas, O., McKemey, K. and Scott, N. (2005), *The Economic Impact of Telecommunications on Rural Livelihoods and Poverty Reduction: a study of rural communities in India (Gujarat), Mozambique and Tanzania*, June, London: Commonwealth Telecommunications Organisation for the UK Department of International Development.

Chapter 3
Innovation System Approach Towards Economic Transformation in the Digital Age

Angathevar Baskaran & Mammo Muchie

Introduction

Until the early 1990s computer and telecommunications (mobile) technologies were viewed narrowly as new opportunities for industrial and export growth. With the onset of Internet Age from the mid 1990s, driven by the new generation information technology that combined computers and telecommunications, increasingly ICT is being perceived by the national governments and international organisations as a vehicle for not only economic growth (in terms of exports and employment), but also for achieving wider socio-economic development which would benefit all in the society, that is they increasingly emphasise on the concept of ICT for development (ICT4D).

Over the years ICT has been increasingly linked to development by academics and policy makers at various levels -- regional, national and international. Particularly, international organizations such as the World Bank and UNESCO have been advocating the use of ICT for achieving broader development. World Bank (2003) made a specific effort to link the ICT projects to the Millennium Development Goals (MDGs). It aimed at stimulating the use of ICT as a tool to reach the MDGs.

The Millennium Declaration, adopted by 189 members of the United Nations in 2000, set 8 Millennium Development Goals (MDG) to achieve improvement in people's lives by 2015, including poverty reduction, education, gender, health and the environment. To achieve this it to set out 18 targets, of which Target 18 calls on the member nations to "make available the benefits of new technologies, specifically information and communications" to the people (ITU, 2004:71). "The Millennium Declaration acknowledges that ICTs are an important tool to achieve its overall goals; ICTs can help alleviate poverty, improve the delivery of education and health care, make government services more accessible, and much more" (Ibid, 2004:71).The SDGs to be realised make science, technology and innovation critical of the 17 goals and 169 targets.

Change in the Perception of ICT

The G7/G8 Summit held in Okinawa in 2000 described ICT as "one of the most potent forces in shaping the Twenty-first Century…fast becoming a vital engine of growth for the world economy" (Ibid). It is argued that "ICT may be applied to almost every problem in probably all sectors… the highest social application of ICT is *poverty alleviation,* since it is the most pressing problem confronting the society" (Ibid). This shift in global perception is illustrated by Figure 1.

Figure 1: Change in the Perception of ICT -- From a Technology for Industrial Growth to Tool for Achieving Overall Socio-economic Development

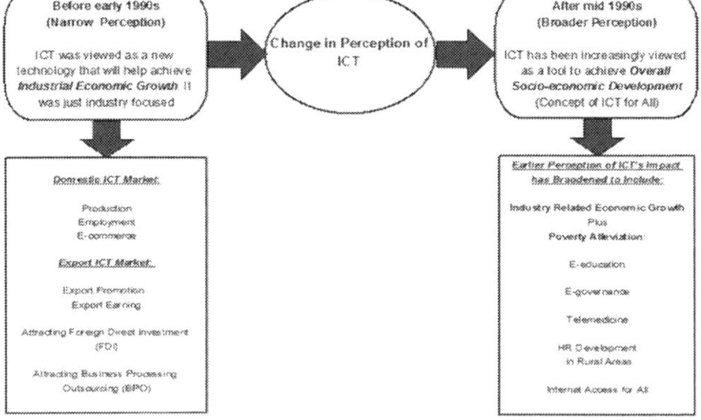

Source: Baskaran, A. and Muchie, M. (2006), *Bridging the Digital Divide: Innovation Systems for ICT in Brazil, China, India, Thailand and Southern Africa*, London: Adonis-Abbey.

World Bank (2005) has put great emphasis on making progress in four of the five development indicators, by employing ICT (economic opportunities, political freedoms, social facilities and transparency guarantees) identified by Sen (1999, p.xii). Similarly, UNESCO (2005) stressed on fostering 'knowledge societies' through ICT to achieve rapid development.

The United Nations Development Programme (UNDP) also believes that ICT could help developing countries to leapfrog into the developed world. Other international organisations such as the World Bank, the International Telecommunications Union (ITU) are involved in plans to promote the spread of ICT by developing an affordable and accessible information infrastructure. Various UN agencies have been promoting the application of ICT by providing assistance to a National Information Infrastructure (NII) and to facilitate access to a Global Information Infrastructure (GII) (Bajwa, 2003:50). Figure 1 illustrates how ICT that was initially perceived as new technology to achieve industrial growth has eventually been viewed as the vehicle for all round socio-economic development by some academics, policy makers, national governments and regional and international organisations. While the governments and international organisations believe that ICT could lead to socio-economic development, they are also concerned over the potential of ICT towards widening the gap between the information-rich and the information-poor, known as the 'digital divide'. It is argued that 'digital divide' could have serious consequences, if governments fail to take appropriate policy measures. Most nations in the developing and developed world are treating this issue seriously and have formulated national ICT policies and initiatives.

The rapid technological change in ICT, particularly since mid-1990s, has enabled the flow of information at a rate and scale that was never witnessed before. This flow of information not only created great opportunity for firms to conduct trade and business on a global scale, but it also created new opportunities and challenges for states to engage in promoting good/democratic governance and social wellbeing. It broke certain traditional distinctions and barriers between nations such as developed and developing world. Unlike the past technological changes, which mainly benefited developed countries; increasing spread of ICT has opened up new opportunities for developing countries to enable the achievement of social and economic benefits. In reality, its potential applications for sustained development and good governance in terms of providing good and efficient public services, and improving quality of people's life have been widely recognised.

Policy makers at national and international levels believe that ICT provide a unique opportunity for particularly developing countries to address problems in the field of education, health, rural development, poverty alleviation and employment. It is also perceived to be a major facilitator for information transparency, good governance, empowerment, participatory management and grass-root democracy. As empowerment requires complete transparency in functioning of government as well as corporations, ICT has an important role to make this possible through extensive use of electronic governance.

Already ICT has made a significant impact in many areas such as manufacturing, productivity gains, public administration, transport, health, education, agriculture, and environment in a large number of countries. At the same time, many countries appear conscious that benefits of ICT could remain confined to some sections of society and creating a new division in society, the so called 'digital divide'. The argument is that if countries failed to foster and manage the ICT revolution properly, it is likely to result in new forms of socio-economic disparity and exclusion by virtue of age, class, gender, religion, ethnicity, language, or illiteracy in both industrialised and developing world. Already, there are indications that the diffusion of ICT in the developing countries is very uneven and it is having some undesirable impacts. While ICT revolution has empowered some groups of people in the developing countries particularly, others are being left out or disempowered (Mansell and Wehn, 1998). This awareness of vast opportunities and challenges posed by the technological change driven by ICT has made many countries both in the industrialised and developing world to formulate strategies not only for building competencies to manage rapid technological changes in ICT but also to ensure its wider diffusion to benefit all in the society.

Information and Computing Technologies for Development (ICT4D)

Review of literature revealed that over the years various conceptual frameworks for ICT for development (ICT4D) have emerged. For example, Thompson (2008) called for a more committed engagement between ICT practitioners and the development community. Hilty and Hercheui (2010) analysed various views and conceptual frameworks related to ICT and sustainable development, such as optimistic and a pessimistic view of ICT with regard to sustainability, the three-pillar approach to sustainable development, the three-level approach to ICT impacts, the claim of human, social and ecological compatibility of ICT, and ICT for development. They identified the problems and limitations of each of these concepts and highlighted the challenges of finding an analytical approach which will effectively support decision-makers in using ICT for achieving sustainable development. They have identified the following challenges: (i) decomposition of the concept of sustainable development; (ii) classification of ICT interactions with aspects of sustainable development; (iii) methods to quantify effects of the ICT life cycle, of ICT applications; (iv) the way ICT changes societal structures (including the economic system); and (v) how to embed these in an organizational or political context of decision making and in a societal discourse (Hilty and Hercheui, 2010, p.233).

Andrade and Urquhart (2010) employed actor network theory (ANT) to examine the different ICT initiatives intended to bring development to underserved rural communities in the Peruvian Andes by providing access to computers and the internet. They found that when interests of the actors involved are not aligned and the network procedures defined by the ICT4D initiative sponsors are unfamiliar to local people, the network cannot be established.

Hayes and Chris Westrup (2012) argue for a rethinking of the notion of context in the ICT for development (ICT4D) research domain. They propose three principles for this: (i) Representation of context is a relational process in which both 'macro' and 'micro' actors engage, and understanding the context of ICT4D requires multiple accounts drawn from different actors; (ii) While research needs to describe the concepts used by 'macro' actors to represent context, it is crucial to recognize them as outcomes of processes of development and not as self-evident entities; (iii) Research needs to attend to how these accounts are produced by 'macro' and 'micro' actors. They applied these principles to the case of M-PESA, an innovative mobile banking application from Kenya and concluded that ICT4D projects in general, and M-PESA in particular, are often difficult to replicate successfully.

Using the case of the rural e-Services project (India), Dearden and Rizvi (2009) argued that as development is a social phenomenon and ICT is a technical phenomenon, ICT for development is inescapably a socio-technical phenomenon. They further argued that ICT design efforts that frame themselves with development objectives require an analysis of their intervention strategies in explicitly socio-technical terms.

Avgerou (2010) argued that that research on ICT and development (ICT4D) involves assumptions concerning the nature of ICT innovation and the way such innovation contributes to development. From the multidisciplinary literature on ICT4D, the author identified two perspectives regarding the nature of the ICT innovation process in developing countries: (i) as transfer and diffusion; and (ii) as socially embedded action; and also identified two perspectives on how ICT contributes towards development:

(i) progressive transformation; and (ii) disruptive transformation. The author concluded that although ICT4D research developed significant theoretical capabilities to study technology innovation in relation to socioeconomic context, it has not provided robust frameworks for arguments about IT-enabled socioeconomic development.

Heeks (2010) argued that the absence or poor quality of assessment of ICT on development was mainly due to lack of conceptual foundations. The author has drawn from models of enterprise value chains, empowerment and Sen's capability theory to demonstrate ICTs' contribution to conceptions of development as economic growth, as sustainable livelihoods, and as freedom.

Heeks (2008) argued that the phase change from *information and communication technologies for international development (ICT4D) 1.0 to ICT4D 2.0* presents opportunities for informatics professionals and offers new markets for ICT companies. Heeks advocated giving priority to ICT application for the poor in developing countries for three reasons: (i) there is a moral argument; the potential benefits of applying new technology to serious global problems such as climate change, civil conflict, disease, resource depletion, and poverty; (ii) , as the poor get richer, they buy more of the goods and services that industrialized countries produce, ensuring a benefit to all from poverty reduction; and (iii) designing a system for an African or Asian community will be more interesting and satisfying. Heeks (2010a) further argued that tools for a digital economy will increasing be shifting towards the world's poor, and the perception of them also will shift from seeing them as victims to seeing them first as consumers, then producers, then innovators of a digital age. That is, Heeks argued that views on the processes and structures of socioeconomic development will change: from Development 1.0 to Development 2.0.

Gomez and Pather (2012) explored the experiences of ICT evaluation in the broad business environment and have drawn parallels with the ICT4D environment. They argued that there is a need for a fundamental paradigm shift in ICT4D evaluation. We argue that it is not sufficient to focus on the easily measurable tangible and quantifiable benefits of ICT, as intangible benefits of ICT4D such as empowerment, self-esteem and social cohesion are more important from a developmental perspective.

Gigler (2011) explored the conditions that can help ICT to enhance the well-being of poor communities, by employing an alternative evaluation framework (AEF) that applied Sen‘s capability approach. Based on the experiences of rural communities in Bolivia, the author concluded that enhancing people‘s informational capabilities, like literacy, is the most critical factor determining the impact of ICTs on their well-being.

Considering the varieties of perceptions of ICT for development and economic transformation, we propose an innovation system approach towards economic transformation in the Digital Age.

System of Innovation towards Economic Transformation in the Digital Age: A Conceptual Framework

The type of national innovation system in a given country provides the context to implement the ICT policies and programmes in both the developed and developing countries. Government policy framework, system of interactions of various actors, institutions and incentives are critical for broadening the application and diffusion of ICT to increase not only industrial economic growth but also to address some social-economic problems such as imbalances in healthcare, education, rural poverty, regional disparity, and so on. Different countries may have differing innovation systems that respond to and deal with different challenges and environments. Hence, it is important to find out how the national innovation systems of different countries influence differently the ICT revolution that is a global phenomenon with a number of associated elements outside the national governance framework. Conversely, it is also important to understand how ICT influences the shaping of systems of innovation in given economies.

The system of innovation is a concept utilised to describe the relationship between internal processes in firms and external processes in the wider environment in the context of knowledge creation, diffusion, and transfer, for example, ICT firms within the context of the global market/economy. There are many types of ***interactions*** that take place both within firms and outside firms. Among the various elements the concept of innovation system identifies is related to ***variation*** of the ***elements*** that constitute ***parts*** of a ***system***. A system of innovation is a concept to distinguish the most significant interactions from less significant ones in relation to actors, activities, and institutions in the process of knowledge creation, diffusion and transfer.

As a *heuristic* concept 'system of innovation' helps to focus on knowledge and learning activities among various actors and institutions that provide competitive advantage in the long-term. The main characteristic of this concept is its flexibility in terms of its application to various geographic regions as well as various organisations. In other words, the type of activity could change (simple or complex, small or big). The actors could change (small firm or large firm); institutions could vary (local or global, public or private). The space could vary (local or national, or regional or global).

The sector/industry could vary. The concept of innovation relates knowledge creation, diffusion and transfer to the actors, activities, institutions, spaces and their interactions. The interactions in the process of acquisition, diffusion and transfer of knowledge can form different degrees and levels of systemic properties and functions around the creation of innovation as the core. The concept of innovation system captures the specific interaction of 'innovation-knowledge' as the independent variable and spatial, sectoral and institutional arrangements as the dependent variable in the context of the activities and actions to bring about transformation and development.

A key attribute of the innovation system concept is the focus it provides in framing problems and the value it signifies to the domain of reality that mainstream economics neglects or underemphasises. The innovation system concept makes central, institutions, histories, territories, technologies, organisations and nations that are often neglected and treated as a residual in mainstream neo-classical economics. The concept has evolved by putting innovation and learning at the heart of the economics of development. The issues of relevance in systems of innovation do not often lend themselves to a reduction in order to fit single disciplinary boundaries. The innovation system is interdisciplinary.

Friedrich List (1856) and his concept 'national production system' may be seen as the historical origin of the national system of innovation (Freeman 1995). Perhaps he is the first to use the system of national economy of production. The innovation system concept has evolved over the years since List's time and its application has been evolving to different sets of problems and areas. Some analysts who work with systems of innovation draw affinities to it with the French Regulation School, and theories of evolutionary and institutional economics in the tradition of Schumpeter (1934) and Veblen (1919).

Innovation system has varied definitions. Of the influential definition, we mention the following: Innovation system has been defined as" the network of institutions in the public and private sectors whose activities and interactions initiate, import, and diffuse technologies" (Freeman, 1987:1). There has been an accent and emphasis on organisations that support R & D and promote the creation and diffusion of knowledge as the main sources of innovation (Nelson and Rosenberg, 1993: 5). Some stress: "All important economic, social, political, organisational, institutional and other factors that influence the development, diffusion and use of innovations" (Edquist, 1997:14).

According to Bengt-Åke Lundvall, the modern version of the concept appeared first in an unpublished contribution to OECD by Freeman (1982) and some years later Lundvall (1985) used the concept in formulating the importance for innovation of the concept of producer-user interaction and feedback for learning. Freeman used the innovation system concept in 1987 when he analysed Japanese economic performance (Freeman, 1987). Subsequently, Lundvall (1992) and Richard Nelson (1993) edited two books on the system of innovation that has become standard references on the subject. Recently attempts have been made to apply the concept for economies in the developing world in general including Africa in particular (Muchie *et al*, 2003).

The innovation system framework suggests that theoretically development is not only a process of *production*, but also it is a process of *innovation*. In addition developing economies are not mere *users* of technology, but also their development should be conceptualised on the domain of active generators or *producers* of technology. Developing economies are not therefore passive recipients of technology from the industrialised economies. The assumption that development can be engineered or steered by technologies that can be transferred from those who control knowledge through intellectual property is a seriously inadequate proposition. There is also the problem that technology transfer cannot be assumed to take place easily. Often the proprietors of technology control knowledge, they do no readily spread it. This provides an additional reason for organising a national system of innovation. Thus there is no alternative that those who wish to use technology must be prepared also to learn to produce it. A necessary condition for development from an innovation system perspective is that those who wish to embark on a sustainable developmental trajectory cannot afford to shy away from trying to develop by being producers of technology, and not stay merely as diffusers, users, absorbers and implementers of technologies produced elsewhere.

For example, the group of economies called the BRICS (Brazil, Russia, India, China and South Africa) are both producers and absorbers, creators and diffusers, though they may be more diffusers and absorbers than creators and producers of technology at present time. One important reason for the taxonomy of the BRICS group is this relative achievement by these economies to strive to produce new technology industries such as, for example an ICT industrial sector. Of course both producers and users of technology undertake different types of innovation. The broad classification of economies as those who are the technology producers as the *innovators*, those that are absorbers of created technologies elsewhere as the *diffusers,* and others like in a n umber of African countries as the *excluded* from being either innovators or diffusers fails to recognise that there is much innovation in being able to absorb technologies as well.

When it comes to the technology specific sector of ICT, it seems to be spreading across the globe cutting across all divisions or barriers. For example, mobile cellular networks are said to have brought 80 per cent of the world's population within reach of a telephone (ITU Database: see Website A).

Since early 1980s the notion of 'technological capability' has been given importance in the development literature. Building technological capability requires deliberate effort and is neither automatic, effortless nor without cost (Bell and Pavitt, 1993:201; Nelson, 1990:77; Dosi *et al.*, 1988:463). The experience of developed economies shows that technological accumulation depends not only on the technological innovations of firms, but also on institutional innovations, which are influenced by social, economic and political factors (Freeman, 1992:170). This explains "why the national system of innovation has worked better in certain environments than in others" (Katz, 1994:251). That is, the rate and shape of technological accumulation varies across countries as they are influenced by the nature of innovation system in their particular country (Nelson and Rosenberg, 1993:4). Also, it is argued that the development of technological capability is relative and area specific, that is, it can be more or less advanced in different sectors or technologies (Enos, 1991:167). Therefore, the effort to build competence and technology accumulation in the ICT sector in different countries is likely to be influenced by their respective national innovation systems. That is, the supporting institutions such as school and higher education system, resource allocation and financial system, overall S&T infrastructure, and the macro-economic environment fostered by a national innovation system provides the basis to foster a technology specific system such as ICT. In other words, the 'whole' (NIS) influences the 'parts' (technology or sector specific systems) and the 'parts' will similarly have an influence on the characteristics of the 'whole'.

A specific technology, left to itself or on its own is neither good nor bad. What is important is how a technology is designed, developed, employed, and whether it can lead to wider social benefits. For example, ICT is believed to have the potential to transform socio-economic conditions of the people. But on its own ICT is not going to bring about socio-economic transformation. It requires appropriate institutional, social, political, and cultural environment to accomplish this. This environment is created by the national innovation system. In turn, national innovation system is shaped by interaction of the physical, human, financial and natural capitals; the interaction of institutions, incentives and knowledge-innovations; and the interactions of economic, social, cultural and political dimensions of development. National innovation system, we would argue, is not just a tool to achieve the narrow goal of industrial/economic competitiveness, but it is about achieving a broader development and wider social benefits.

It is quite possible for a country to develop a strong and highly efficient technology specific innovation system, even if it has failed to create an equally efficient national innovation system. In such circumstance, the experience of technology specific system could benefit the national innovation system. The Indian space technology specific innovation system is an example for this (Baskaran, 2001 and 2001). In the case of ICT, as it is distinct as a technology enabler with pervasive influence on different industrial and social sectors, it has a very significant potential to influence the characteristics of the national innovation system. For example, other technology specific innovation systems such as biotechnology are unlikely to have similar impact on different industrial and social sectors. Therefore, we argue that particularly the ICT specific innovation system is likely to draw from and as well as benefit the national innovation system.

The linkage between the national system of innovation, the ICT specific innovation system, and (narrow) industrial-economic and (broader) social-economic development objectives is conceptualised in the Figure 2. It attempts to capture the complex relationship between these two systems through simple illustration.

Figure 2: Linkages Between Systems of Innovation and Industrial-economic/ Socio-economic Objectives

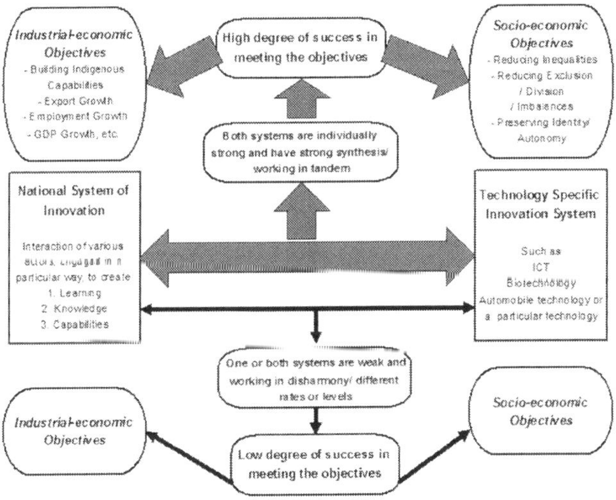

Source: Baskaran, A. and Muchie, M. (2006), *Bridging the Digital Divide: Innovation Systems for ICT in Brazil, China, India, Thailand and Southern Africa*, London: Adonis-Abbey

Figure 2 presents a number of possible scenarios. On the extreme positive side, the best will be where the general national system of innovation and the ICT specific innovation system are strong and show a high degree system synergy, synthesis and work in tandem. This results in greater success in meeting both industrial related economic growth and broader social-economic objectives. In such environment, ICT can play a very positive role. On the extreme negative side, the synthesis between the broader national system of innovation and the ICT specific innovation system is weak and they work in disharmony, moving at different rates and levels, failing to connect, and achieve system synergy and dynamics. In this case, both national innovation system and ICT will not enable the fulfilment of either narrow or wider social-economic objectives. Instead of positive contribution, ICT can play a negative role. Between these two extremes, it is possible there are varying degrees of system synergy and synthesis between national system of innovation and the ICT specific innovation system. If this is relatively high, then some industrial and social objectives may be fulfilled but not all. If this is low, only few objectives can be realised.

Various elements involved in creating an efficient or lopsided innovation system for ICT is illustrated by the Figure 3.

Figure 3: Agents Involved in an ICT Innovation System

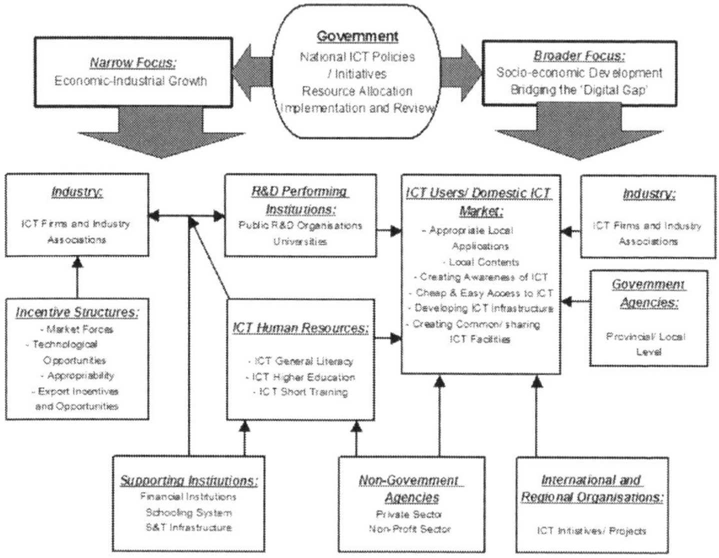

Source: Baskaran, A. and Muchie, M. (2006), *Bridging the Digital Divide: Innovation Systems for ICT in Brazil, China, India, Thailand and Southern Africa*. London: Adonis-Abbey.

In the area of economic industrial growth through ICT (the narrow objective) the major elements that play instrumental role are the industry, the R&D performing institutions, the incentive structures, availability of ICT human resources and the supporting institutions. They interact with each other and determine the rate and shape of capability building in ICT. In the area of socio-economic development and bridging the 'digital divide' (broader objective), the elements that play major role are the local and provincial government agencies, non-governmental agencies, R&D organisations, industry, ICT human resources, easy and cheap access to ICT, local involvement and awareness, and development of ICT infrastructure at the local level (rural areas). When the inter linkages are stronger between various elements, then it leads to an efficient ICT innovation system. When the linkages between them are weaker or when some actors are absent (or not strong), then it is likely to lead to an uneven or lopsided ICT innovation system.

Conclusion

The importance of creating a dynamic ICT sector and integrating this sector to the wider system of innovation can create significant developmental economies, efficiencies and dynamics. It is thus pivotal that ICT is recognised as a key technology system that can strengthen a country's national system of innovation. Its importance in enabling to contribute to economic growth and solve social issues is also clear. But like any thing useful, it must be utilised with goals set by clear social policy imperatives, if its value is to be helpful rather than negative.

Unless development of ICT is also shaped within a socially and politically framed national system of innovation, it is likely to create more diseconomies than economies. ICT would not be as productive as one would hope for, if there is no value addition in knowledge and learning that accrues from the linkages between ICT and the economy. Where the value added from the ICT would come would be when it is integrated with the national innovation system to achieve what Frederick List (1789-1846) called mental capital building as a means of wealth accumulation. Unfortunately, it appears that in many poor countries ICT will be used mainly to accelerate the primary export economy and not to transform it, as many of the poorer economies do not even have rudimentary national system of innovation. Thus, less developing economies face twin challenge of creating a national innovation system to foster the institutional pillars of national innovation system and developing and integrating an ICT system with the former. The synergy between these can provide the basis for a developmental dynamic in the poorer developing economies.

However, it is evident from the case studies that while we cannot ignore the potential of ICT to generate wider socio-economic benefits, we should not fail to recognize the limitations of ICT in solving social problems, as they will not be solved by technology alone.

References

Andrade, A. D. and Urquhart, C. (2010), 'The affordances of actor network theory in ICT for development research,' in *Information Technology & People*, 23(4): 352-374.

Avgerou, C. (2010), 'Discourses on ICT and Development', in *Information Technologies & International Development*, 6 (3): 1-18.

Bajwa, G. S. (2003), 'ICT policy in India in the era of liberalization: its impact and consequences', in *GBER*, 3(2): 49-61.

Baskaran, A. (2000), 'Duality in national innovation system: the case of India', in *Science and Public Policy*, 27(5) 367-374.

Baskaran, A. (2001), 'Competence building in complex systems in the developing countries: the case of satellite building in India', in *Technovation*, 21 (2): 109-121.

Bell, M. and Pavitt, K. (1993), 'Technological accumulation and industrial growth: contrasts between developed and developing countries', in *Industrial and Corporate Change*, 2(2):157-210.

Dearden, A. and Rizvi, H. (2009), 'A deeply embedded sociotechnical strategy for designing ICT for development', in *International journal of sociotechnology and knowledge development*, 1 (4): 52-70.

Dosi, G., Freeman, C., Nelson, R. and Soete L. (ed.) (1988), *Technical change and economic theory,* London: Pinter Publishers.

Edquist, C. (ed.) (1997), *Systems of Innovation: Technologies, Institutions and Organisations,* London: Pinter publishers.

Enos, J.L. (1991), *The creation of technological capability in developing countries,* London: Pinter publishers.

Flor, A. G. (2001), 'ICT and Poverty: The Indisputable Link', Paper presented to Third Asia Development Forum on "Regional Economic Cooperation in Asia and the Pacific", Organized by Asian Development Bank, 11-14 June, Bangkok.

Freeman, C. (1982), 'Technological infrastructure and international competitiveness', Draft paper submitted to the OECD Ad hoc-group on Science, Technology and Competitiveness, August, mimeo.

Freeman, C. (1987), *Technology Policy and Economic Performance: Lessons from Japan,* London, Pinter publishers.

Freeman, C. (1992), 'From scientific and technical institutions in the national system of innovation', in Lundvall, B. (ed), *National systems of innovation: towards a theory of innovation and interactive learning*, London: Pinter Publishers. Freeman, C. (1995), 'The National Innovation Systems in historical perspective', in *Cambridge Journal of Economics*, 19 (1).

Gigler, B-S. (2011), 'Informational Capabilities'- The Missing Link for the Impact of ICT on development', *Working Paper Series No. 1*, March, 2011, E-Transform Knowledge Platform Working Paper, World Bank, Washington DC. Available at: http://siteresources.worldbank.org/INFORMATIONANDCOMMUNICATIONANDTECHNOLOGIES/Resources/InformationalCapabilitiesWorkingPaper_Gigler.pdf

Gomez, R. and Pather, S. (2012), 'ICT Evaluation: Are we asking the right questions?', in *The Electronic Journal on Information Systems in Developing Countries*, 50(5): 1-14.

Hayes, N. and Westrup, C. (2012), 'Context and the processes of ICT for development', in *Information and Organization*, 22: 23-36.

Heeks, R. (2008), 'ICT4D 2.0: The Next Phase of Applying ICT for International Development', in *Computer*, IEEE Computer Society, June 2008.

Heeks, R. (2010), 'Do Information and Communication Technologies (ICTs) contribute to development?', in *Journal of International Development*, 22: 625-640.

Heeks, R. (2010a), 'Development 2.0: Transformative ICTEnabled Development Models and Impacts', *Development Informatics Short Paper no.11*, Centre for Development Informatics IDPM, SED, University of Manchester, UK. Available at: http://www.seed.manchester.ac.uk/medialibrary/IDPM/working_papers/di/di_sp11.pdf

Hilty, L. M. and Hercheui, M. D. (2010), ' ICT and Sustainable Development', In: *What Kind of Information Society? Governance, Virtuality, Surveillance, Sustainability, Resilience, IFIP Advances in Information and Communication Technology*, Volume 328/2010, Berleur, J., Hercheui, M.D. Hilty, L. M. (eds.), Berlin Heidelberg, New York: Springer, pp. 227-235.

International Telecommunications Union (ITU) (2004), *World Telecommunications Report 2003,* Geneva: ITU.

Katz, J.M. (1994), 'Technology, economics, and late industrialisation', in Salomon, J. (ed), *The uncertain quest: science, technology and development*, Tokyo: UN University Press.

List, F. (1856), *National System of Political Economy,* Philadelphia: Lippincott.

Lundvall, B. A, Johnson, B., Andersen, E.S. and Dalum B. (2002), 'National Systems of Production, Innovation and Competence Building', in *Research Policy*, 31(2):213-31.

Lundvall, B. A. (1988), 'Innovations as in Interactive Process: From User-Producer Interaction to the National System of Innovation,' in Dosi, G., *Technical Change and Economic Theory*, London: Printer.

Lundvall, B. A. (ed.) (1992), *National Systems of Innovation: towards a theory* of *innovation and interactive learning*, London: Pinter.

Mansel, R. and Wehn, U. (ed.) (1998), *Knowledge Societies: Information technology for sustainable development* , Oxford: Oxford University Press.

Muchie, M., Gammeltoft, P. and Lundvall, B. A. (2003), *Putting Africa First: The Making of African Innovation Systems*, Aaalborg: Aaalborg University Press.

Nelson, R R. (1990), 'On technological capabilities and their acquisition', in Evenson, R.E., and Ranis, G. (ed)., *Science and Technology: lessons for development policy*, London: Intermediate Technology Publications.

Nelson, R. (1993), *National Innovation Systems a Comparative Analysis*, New York and Oxford: Oxford University Press.

Nelson, R. and Winter, S. (1982), *An Evolutionary Theory of Economic Change*, Cambridge: Belknap Press.

Nelson, R. R. And Rosenberg, N. (1993), 'Technical innovation and national systems', in Nelson, R. R. (ed), *National innovation systems*, New York: Oxford University Press.

Nelson, R.R. (1987), *Understanding Technical Change as an Evolutionary Process*, Amsterdam: North Holland.

Schumpeter, J. (1943), *Capitalism, Socialism and Democracy*, New York: Harper.

Thompson, M. (2008), 'ICT and Development studies: Towards development 2.0', in *Journal of International Development*, 20: 821–835.

UNESCO (2005), 'Towards Knowledge Societies', *UNESCO World Report*, Paris: UNESCO Publishing.

Veblen, T. (1919), *The Place of Science in Modern Civilization and Other Essays*, New York: Huebsch.

Website A: ITU World Telecommunications Indicator Database. Available at: www.itu.int/publications/folderdetails.aspx?lang=e&folder=D-IND-WTID-2004&menu=categories.

World Bank (2003), *ICT for Development: Contributing to the Millennium Development Goals: Lessons Learned from Seventeen infoDev Projects*, Information for Development Program, World Bank, Washington DC.

World Bank (2005), 'Financing Information and Communication Infrastructure Needs in the Developing World: Public and Private Roles', *World Bank Working Paper 65*, Global ICT Department, World Bank, Washington.

Chapter 4
Building the Information Society for Us All

Christopher Freeman & Luc Soete

1. Introduction: The Vision

The 1990's have witnessed a great proliferation of reports and papers on the Information and Communication Technologies (ICTs) and on what is now described as the 'Information Society' (IS). This burgeoning literature is mainly concerned with the advent of the 'Information Highways' and the ways in which even greater amounts of information may be distributed ever more quickly, efficiently and cheaply to more and more households and businesses throughout the world.

The objective of this report is not to add one more document to an already over-subscribed field. Rather, it is to transcend this discussion from the outset, by making a separation between data and information, and by distinguishing between the notion of information and the need for knowledge. It is necessary to separate out the transmission of data from communication between people, and the acquisition of knowledge. Most present-day telecommunication systems are still systems of transmission of data. In this report, we focus our attention on ways in which information can be converted into useful knowledge, so that the 'information economy' may become a 'knowledge-based economy' and the technology be put at the disposal of all. The underlying ICTs provide the potential for great increases in productivity and many new and improved products and services (see amongst others Freeman and Soete, 1990, 1994). However, history shows that the ability to convert this potential into actual gains in productivity and improvements in living standards and quality of life depends on a prolonged process of learning and institutional change (see amongst others Freeman and Foray, 1993).

The technology in itself is neither good nor bad. It is the use which human beings make of any technology which determines both the nature and extent of the benefits[1]. Moreover, these do not accrue automatically to everyone in society. For most innovations, both benefits and costs are unevenly distributed. While some individuals and groups may benefit greatly, others may be seriously disadvantaged, through for example, loss of employment or erosion of skills.

We welcome the opportunity to present this interim report hoping that it might provoke discussion. [...] It gives us the chance to identify a number of key themes which we believe should be moved to the centre stage of the discussion of the IS, realising that it is neither possible nor desirable to separate the social and societal aspects from the technical, industrial and economic ones.

In the future there could be different models of Information Societies, just as today we have different models of industrialised societies. They differ in the degree in which they avoid social exclusion and create new opportunities for the disadvantaged. In the White Paper on Growth, Competitiveness and Employment, the European Commission underlined the importance of the social dimension which characterizes the European Model. A strong ethos of solidarity should also characterize the European Model of the Information Society. This is not an easy goal to achieve since the traditional structures of the welfare state will have to undergo substantial changes. To adapt to these changes, an active rather than passive concept of solidarity is needed. We would like to emphasise four features of such an active solidarity.

First and foremost, it is essential to view the IS as a "learning society". The issues of education and training were given to us to consider as a theme of our work. We agree that education and training is an important sector, but the sectoral boundaries cut across the formal provision of education and training services. For us, learning and the development of knowledge should be a lifelong process which takes place at work and in the home, as well as through education and training. Furthermore, putting the central emphasis on the learning society, rather than the IS per se, provides a potentially much more positive route forward for the IS debate. In particular, the more appropriate emphasis on learning illustrates that, although access to the information super-highway is and will continue to be important, it is insufficient to ensure the rapid development of a knowledge-based economy. As we noted above, it is the ability to use information and to transform it into knowledge that is important. This is why we argue below in favour of an approach to access to information which goes beyond conventional ideas about universal service provision (USP).

Second, as information is assuming a much more important position in our lives, a crucial additional characteristic of a human information society is undoubtedly that people should be in control of the information, rather than it being used to control them. In practice, with the proliferation of detailed information recording our movements, purchases and personal profiles, there is a risk that an intrusive IS will develop. We would like to see this risk reduced. Nevertheless, we would still like to draw attention to the trend towards individualisation, which underlies these powerful new technologies for mapping and monitoring activities. The collection of this wide array of information concerning the activities of individuals allows the customisation of all sorts of areas of everyday life, down to the individual level. There are observable trends towards the individualisation of various aspects of life: individual contracts of employment, targeted benefits, consumer profiling, personalised health care, individualised insurance planning and so on. There are obvious advantages in terms of effectiveness of such systems in that, for example, it will be possible to make services really responsive to individual needs -- but there are also major risks in terms of the invasiveness of such systems and the likely reductions in social solidarity (see amongst others Freeman, 1991).

Third, human activities (work, leisure, contacts with administrations, banks, etc.) will occur more and more through the intermediary of telecommunication networks. These activities will be increasingly based on representations of reality (that is, abstract and virtual images of reality), rather than reality itself. There are significant advantages to this evolution such as, for example, (faster delivery of products and services, fewer accidents and, perhaps, lower physical stress), but there are also risks. Virtual life is not real life and the representation of reality is not reality. Some of us will not be able to endure living permanently in a representation of reality.

Fourth, in any period of major social, economic and technological change, the balance of social costs and advantages shifts. The Information Society is no exception. Changes in the industrial structure are leading to job destruction, as well as new job creation. There are great challenges in terms of adjusting to the IS at the sectoral level, at the firm level, at the level of individual skills and occupations, and within both public and private service provision. Furthermore, there are similarly dramatic changes in the organisation and quality of work. Organisations have to be flexible to perform in the new paradigm of competition of quality, customisation and rapid innovation. Workers are increasingly required to be flexible in terms of the skills they have, the tasks they perform, and the new forms and structures of employment. New relationships between paid employment, work and activity are emerging. These raise questions about the place of work in people's lives, particularly for the large numbers of people faced with unemployment, under-employment, or unstable employment patterns (see amongst others Freeman and Perez, 1988; Freeman and Soete, 1994; Freeman, Soete and Efendioglu, 1995).

The Information Society is emerging at a time of great change. Without any doubt, new ICTs provide tremendous opportunities for new growth and employment creation; for a more efficient use of inputs, not just of labour, but also of energy, materials and capital, contributing to a potentially more sustainable development path; for higher income and welfare; for more decentralised organisation forms, whether on the shop floor or in terms of work-sharing; for more consistent regional and urban development patterns; for individual enrichment as well as a more democratic and decentralised society. Importantly, the opportunities for higher economic and productivity growth crucially depend on congruence between the technological, economic and social dimensions (Freeman, 1989).

No outcome is predetermined. Thus, just as in the case of industrial and commercial enterprises, where the adoption of new technologies will be the subject of cautious analysis and will rarely be based on technical efficiency alone, so too should the societal adoption of new technologies be based on policy debate and on the search for measures necessary to achieve an economically and socially integrated IS.

The Information Society is a society in formation. The decisions we make now, and the opportunities we take now, to construct a strong social dimension for the future of Europe may well be with us for a long time. We cannot afford to miss out in the global stakes nor can we afford to weaken the solidarity of Europeans in a rush for a rapid move into the IS for some, but not for all.

Social policy therefore merits equal if not more weight than economic policy in formulating our approach to the IS. We believe that the Commission has paid insufficient attention to these issues so far, and in the following sections we elaborate these arguments in more detail.

2. Employment

There is a large potential for the growth of completely new forms of employment in the IS. New core ICT intensive sectors are forming the basis of new industries and multi-media based industries in particular are likely to be high value, high skill sectors with considerable labour intensity. On a positive note, therefore, we see a potential for new knowledge-rich forms of employment emerging, which could create highly satisfying employment for many Europeans. The prospect of new occupations based upon information services which reinforce creative, collaborative, communal and caring aspects of work are strong and should be an area which policies seek to build up, not only because jobs with these attributes are socially desirable, but also because in many cases these innovative services will form the basis of the emerging information economy in Europe, and thus also a major component of its competitive advantage.

However, we also feel bound to raise the question to what extent new technologies represent a threat to employment, a subject which has been at the centre of polemical debates between economists and sociologists for the last two hundred years (see Freeman, Clark and Soete, 1982; Freeman and Soete, 1987).

While we do not subscribe either from an economic or historical point of view, to such alarmist views as "The End of Work", there are, nevertheless, some new features of the emerging IS which have in our view been insufficiently taken into account in the current employment debate[2].

- New ICTs seem to have a powerful potential to affect service employment, which has been a traditional employment "reservoir" in most industrialised countries and today represents over two thirds of total employment in the EU. Many of these service jobs both in the private and public sector hitherto largely have been "protected" from "labour-saving technical change". To assess the impact of the use of new ICTs on such service employment, we need to know much more about the relationship between innovation, and the creation and destruction of jobs in such sectors (see also Freeman, 1995a).
- The deregulation and privatisation of telecommunication operators, and the increased possibilities for the delocalisation of particular activities, are likely to have major structural impacts on existing manufacturing and services employment. At least in the short run, there is concern that this impact might involve substantial employment displacement.
- Employment statistics are still heavily biased towards manufacturing and material goods, as are our statistical methods for estimating inflation. With the growing proportion of consumption being devoted to "information-intensive" goods and new information services, one may wonder whether current inflation estimates correctly measure overall consumer price trends in the emerging IS. There is at least a suspicion

that the substantial price declines linked to new information goods and services are at present not sufficiently being captured in official inflation estimates. Again, little is known about the size of this issue and even less about its macro-economic policy implications.

- Like many others in this field, we are of course concerned about the fact that the benefits and costs of the use of new ICTs might become more and more unevenly distributed with some groups being seriously disadvantaged by loss of employment and erosion of skill. How to respond to this "distributional" challenge is to some extent the central issue of our report. However, this debate should not be limited to the traditional set of "social policy" proposals. It should also address the erosion by the emerging IS of the underlying financial base of existing "distributional" policies (such as many of the current European social security systems) together with potential and more specifically IS-related "distributional" policies.
- Finally, we share the growing concern that the combined effect of such distributional impacts, and the growing international transparency of markets, will put the burden of adjustment even more on the unskilled labour and low wage parts of the labour force. The emerging information society is a global one (see in particular Freeman, 1994, 1995b, 1996). The challenge it is likely to put to all our institutions is a formidable one.

3. Work Organisation

The relationship between work organisation and the information society is complex. Whilst new network technologies are major factors allowing for organisational innovation, there are many other drivers of change, such as increased demands for quality, global competition, and market volatility. ICTs can be seen as a means of achieving the productivity and performance improvements necessary to control costs and retain competitiveness.

- The emergence of new forms of organisation which increasingly use networking structures to increase flexibility, seems to be a widespread feature of these trends. There is, however, a considerable amount of latitude in the way in which the new technologies are introduced. Not all ways of implementing ICTs are equally effective. There is much evidence to suggest that those approaches to innovation which build upon the social relations which are inherent in the design and implementation of new technologies, are likely to result in higher performance workplaces by comparison to approaches seeking only to achieve performance in terms of the technical or economic efficiencies.

- A key feature of recent changes in organisational structures is the emergence of greater decentralisation as evidenced, for instance, in the clear shift towards outsourcing and externalisation, and downsizing and devolved hierarchies. Such changes provide important challenges to our existing systems in terms both of ensuring high standards in employment, and providing the appropriate incentives for investment in training, especially with regard to people who work in small firms, in services, and in the self-employed and freelance sectors.
- Teleworking (whether it be home-based or in teleworking centres, full-time or part-time) is more than "work at a distance". It is significant not only because it is directly associated with ICTs, but also because it involves new flexible patterns of management and work organisation. We are concerned that teleworking should develop in ways which emphasise the positive aspects of the new flexibilities (like wider access to jobs, more flexible timing of work), and which reduce the risks of shifting away from conventional workplaces such as isolation of workers, lower social protection and inadequate working conditions.

The future of work and social relations [3]

The nature of work and its role in our lives seems to be undergoing major changes. Although the nature and extent of these changes varies markedly from place to place, the general dimensions in a few key areas of change are an increase in part-time work and self-employment; an increase in the unpredictability of working hours; an increase in casualised forms of work (temporary or fixed term contracts, etc.); the increasing participation of women in work; the trend towards earlier retirement and the decline in the expectation of a career for life.

On some levels, the increased use of ICTs both increases the salience of these trends and, in some areas, also provides scope for new policies to improve the integration of working life into the rest of our lives. For instance, the much more rapid rate of obsolescence of existing skills is undoubtedly leading to increasing involuntary early retirement. Similarly, the rhythm of work in the IS will be far less predictable, as people will be expected to work over a more extended working week.

- The reconciliation of work and home life will be a major issue in the social cohesion of the IS. It is necessary that the complex relationships in this area be better understood in order to inform policy making.
- Traditional social relations rest on a clear distinction between place and time at work and place and time off work. The IS tends to re-structure these accepted conventions of place and time such as family

space and time; work space and time; leisure space and time; training; and consumption. The boundaries between these activities, in particular between work and non-work, are blurring. That is why new social relations need urgently to be considered; one must foresee and negotiate the right to go off-line and to be free from the constant call of the electronic network at anytime and in any place.

- Similarly, new remuneration systems will have to be negotiated. With the shift towards more flexible working hours, and indeed more flexible work arrangements more generally, new arrangements to make flexible work more attractive to workers might be found, such as annualized working time contracts, and the use of time off in compensation for unsocial working hours.
- The possibility of home-based teleworking also raises a whole new set of problems concerning the reconciliation of work and home life. The problems of reintegrating work into the home, even on a part-time basis, have significant economic, social and psychological implications.
- Not only will social relations be modified by the questioning of traditional work/non-work frontiers but sectoral frontiers will also be questioned. The new flexible work organisation patterns, internationalisation, and shifts in the sectoral structure of the economy, all create pressures to adapt the machinery of representation and negotiation between the social partners. Thus, we believe that innovation in procedures of social dialogue and collective agreements will be needed in order to maintain the salience of these channels in the new flexible regimes of the IS.

Labour markets

The Information Society is emerging at a time when the traditional expectations of work and career are facing a radical challenge. As work becomes more subject to change in terms of its content (through continual innovation) and in terms of its security of tenure, labour market policies must become more responsive and proactive. There has been a rising interest in active labour market policies across most of Europe in recent years. It would seem that in the IS, such proactivity is likely to become more important, and the possibility of finding new ways to improve the performance of the labour market is enhanced.

- The effectiveness and relevance of external labour market measures will require careful reassessment in the context of the IS, given the changing organisational, occupational and skill profiles of work. Of particular importance are training into work schemes aimed at people with relatively low qualification levels, the long-term unemployed, and other target groups which are more at risk of exclusion in the IS.

- Internal labour markets are likely to be under pressure to become more adaptable, so that organisations are able to meet the challenges of on-going innovation which will be associated with the IS.
- Increases in the efficiency of the labour market, and especially the link between the internal and external labour market, will be required in order to deal with the increasing flexibility of work. ICTs provide at least a partial solution to these problems by increasing the transparency of the labour market, reducing the costs of providing information sources in a wide variety of places and allowing easier access to information on employment demand and supply. However, for such systems to be effective, some new approaches within the labour market institutions will be necessary, if only to safeguard equality of access, security and reliability.

4. Social Cohesion and the Quality of Life[4]

Work and employment are by no means the most important dimensions of a broader social vision of the IS. Any such vision has to take account of the major possibilities for improving the quality of life and social cohesion through creative approaches to ICTs. However, equally there are concerns that the IS could lose its sociability. The culture of screen and chair could lead towards isolation, with people increasingly interacting through the digital media and withdrawing from the social realm.

- A wide diversity of different ways of interacting is desirable so that people have a real choice between interacting on-line and interacting with humans. The IS offers new opportunities for social integration, through building up communities at a local level. The introduction of ICTs will permit more people to work from, or near, home for at least some of the time, meaning that more localities might come alive during the working day. With ICTs, more services, especially education and community services can be delivered into communities at a local level. Such changes could help to reinvigorate whole communities and lead to stronger social networks and a sense of place.
- ICTs could help to overcome some of the disadvantages associated with mobility problems or a lack of access. Indeed, the opportunities associated with the IS to increase the quality of life of disadvantaged groups are already provoking great excitement and attention. Steps

should now be taken to shift from speculation to concrete action.

- The introduction of ICTs could introduce new risks of social exclusion for some groups and exacerbate the risks faced by other groups who are already disadvantaged. People who are not in the workforce or education are less likely to encounter these new technologies, and so may be left behind as the new technologies become more widespread. The people at risk of being left out, because they are not in work include those who have retired early, those taking career breaks to raise children and unemployed people. People with low incomes, including many elderly people, single parents and, again, the unemployed, face particular dangers that they will lack access to the IS, because the household budget will probably not be adequate. The nature of exclusion in the IS needs close scrutiny so that policies can be constructed to avoid the development of a two-tier IS.
- In policy terms, it is important to recognize the need to adapt the IS to the needs of people, and not just to expect people to adapt to the IS. Although there are clear barriers of technical literacy facing most people, it seems pertinent to ask whether the current user-friendly technologies cannot be made a lot friendlier to use through more ergonomic hardware and software.
- There is a fear that the IS could be an isolated society, with human contact increasingly replaced by telepresence and electronic communication. However, the reality seems to be more complex. Leading examples where the automation of services has taken place, such as in retail banking, seem to result in a reduction of routine human transactions but an increase in more complex interactions, such as giving advice, consultation and negotiation. These changes are not yet well understood, nor do we know what can be done to make the new systems of social interaction more convivial.
- The family has an important role as a place where people will be socialised into the IS and as a learning environment. Ways in which families can fulfil these roles more effectively are needed, such as closer integration between learning at home and learning at school and support for parents in understanding the new ICTs, allowing them to learn along with their children.
- The social consequences of the rapid introduction of new ICTs are not well understood. Although the more dramatic scenarios of fragmentation and isolation associated with a cyber-society seem simplistic and overstated, there are concerns that the abstract nature of interacting

through screens could lead to an abstraction of reality. A particular issue is the stress associated with information and perceptual overload. The sheer volume and immediacy of information raises fears that people are no longer in control and need new techniques and strategies to help them cope. In addition, these technologies extend people's availability, so that they are on-line and on-call wherever they are and at all times. Such accessibility could easily become invasive, and its effects on social life need to be better understood.

- New forms of marketing, retailing and consumption are emerging along with the IS. In the past, most retail services were local and available for a few hours of the day. In recent years a widening of the availability of these services has occurred in most places. With ICTs, it is possible to buy goods and do business, on-line, at any time and in any place. These changes will affect the way we consume, our understanding of the process of consumption, the relationships between buyer and seller and the place of shopping in our culture. They need to be tracked to enhance the ability of consumers to use the new technologies and services to their own advantage.

5. Regional Cohesion

ICTs have a much remarked-upon and potentially powerful ability to 'shrink distance'. In essence, new industrial and social geographies could emerge, particularly with regard to services which can be delivered over the wire. In principle, therefore, there are new policy challenges concerning the relative economic advantage of non-core regions, especially the less favoured regions (LFRs) in Europe. In the first place, the fact that the telecommunications infrastructures are relatively less well-developed in (some of) these regions could lead to them losing out on the potential economic competitiveness of being able to provide and access services on-line. This could reinforce existing disadvantages of peripherality, especially in lagging regions, or could simply annihilate the potential advantages of some LFRs in terms of lower labour costs. On the other hand, the potential for catching-up, or even regional 'leapfrogging', could be high in distant regions leading to new commercial activities and the emergence of service providers. The impact of deregulation and liberalisation is likely to make such new opportunities much more transparent.

In essence, therefore, the development of an integrated approach to the IS in order to increase regional cohesion is necessary. This implies a number of interrelated steps.

- There is a need to develop much more focused and targeted approaches to infrastructural support. Following this perspective, Community Structural Funds could be directed in ways which are much more sensitive to emerging demand, and thus more effective and efficient.
- Any reformulation of USP has to take into account new technological possibilities and be accepted by all countries in the European Union. The problem is how to define a level of access which is adequate (especially given the proliferation of media), which will have some long-term validity, and which is operational at a reasonable cost. There is unlikely to be a blanket approach which will be good for all regions, and so any programmes and policies adopted should be attuned to the variations which exist.
- Broader measures are needed, however, which can ensure that different social groups (e.g. residential, schools, hospitals, small enterprises, public services) gain access to a reasonable service level at a reasonable price.
- Satisfying demand for telecommunications services is as important as investment in the infrastructure and services. A critically important matter is the modernisation of enterprises and the public sector, and the development of human resources to increase the development impact of ICTs.
- Regional institutions, such as chambers of commerce, regional innovation centres, collaboration between regional government, and training institutes and enterprises, are important to the innovativeness of the region and its capacity to respond to the challenges of the IS. In LFRs, they are often thinly developed, rigid and not up-to-date, or not well-integrated.

6. Education and Training

We have stressed from the start that the IS should be viewed as a learning society. Education and training is a horizontal issue cutting through most aspects of this report, as well as an area of specific concern to the educational and vocational training sector. The centrality we allocate to learning, of course, means that we immediately emphasise the need to look beyond training in ICT skills. In addition, whilst we would underline the importance of ICTs as new tools, techniques and media for delivering education, we argue against a narrow focus on new learning technologies and systems.

Instead, new and changing skill demands and new possibilities for learning methods should be seen as components of an integrated approach to lifelong learning in the IS. The aim should be to meet social and economic criteria of openness, accessibility, affordability, and, as far as possible, learning should be challenging and fun. Most of all we want to underline the importance of the social nature of the learning environment, whether at school, college, the workplace, or in the home. Much of the value of education resides in learning collectively. Education is valuable, therefore, not just for its instrumental role in developing skills, but also for its role in the development of social cohesion and culture.

- New multi-media software for education and training is a fundamental requirement for the IS. Present software programmes are often not very satisfactory, which partly explains their low levels of adoption. The quality of such software needs to be of the highest standard and has to be customised to the different needs in different training environments around Europe. Thus, a strong European industry producing these systems is necessary to make sure that supply of the training and education is relevant to European needs, and the involvement of trainers in the design of these systems is needed to make sure it really is useful in the training or educational environment.
- Because education and training is mostly a collective experience, a central focus of new learning approaches will be on developing a new role for the teacher as a guide through the information maze and as a coach, rather than a lecturer. New learning approaches will also be needed to promote the lifelong learning of trainers.
- Equality of access to education and training is, again, an area of challenge, and one in which the IS can make a contribution. All communities and social groups will need access to the infrastructure of the IS and the means to make use of the services which are becoming available. On the other hand, ICTs can reach out to people who cannot currently easily obtain access to classrooms to learn, particularly those who live in remote areas or who are disabled and/or housebound.
- ICTs also provide new ways of making education and training available to people for whom attending regular classes is difficult - especially those in employment or with family commitments. Thus, whilst we place great emphasis on the conviviality of the learning environment, ICTs can undoubtedly increase one's chances for gaining access to learning opportunities, through enhanced open and distance learning.

7. Health

As with education and training, there is great scope for innovation in health service provision in the IS. ICTs can directly contribute to improved diagnostic capability and health care and a more cost effective delivery of services. Clearly, in any such innovations, patient safety and the acceptability of these systems to both patients and medical staff are fundamental.

- Given the vast potential benefits of health telematics, we are primarily concerned that these technologies should diffuse rapidly, through the development of the technological systems themselves and by increasing their accessibility to hospitals, doctors and patients (and making sure that there is equality of access); promoting standards for sharing medical information and identifying and removing barriers to their adoption.
- The new technologies will, inevitably, change the jobs and skills required of health service staff. The participation of health care professionals in the design and implementation of these systems will lead to a more rapid and effective adoption of these technologies. In addition, however, there are major training and retraining requirements which should be addressed in parallel to increase the acceptability, performance and cost effectiveness of the new technologies.
- New safeguards will be needed so that both professionals and patients will have confidence in the new systems. The collection and transfer of medical records will occur to a far greater extent than in the past. The confidentiality of health data, therefore, is a major priority. In addition, the detailed recording of information on treatments means that clear guidelines will be needed on the rights of access to this information, and on issues of liability which might arise where there are breakdowns in the systems or disputes over the outcomes of treatment. In some cases, ethical standards in the health sector may need re-examination.
- As with the role of the teacher in education, the physician-patient relationship is the pivotal link in the delivery of health care. Approaches to health care in the IS should enhance the quality of this link by providing more support in terms of diagnostic power and information to the physician. Health care in the IS should, in general, aim for improvements in health, as well as for improvements in the quality of care itself.

8. Culture and Media

Cultural change assumes extraordinary importance in the IS, if only because of the centrality of communication and knowledge in culture. Language, for instance, is one of humanity's greatest cultural technologies.

ICTs are not just a new wave of contemporary machines and industrial processes, but also are systems for recording and communicating information. Thus, the cultural effects of the widespread diffusion of this technology are likely to be especially profound in comparison with earlier waves of technical change and thus figure prominently in our report.

Technology does affect culture and the economy, but there is no question of a uni-directional linear model. Each interacts with the other and so there is the possibility to shape technologies in support of cultural and political objectives in a number of areas.

- The IS can be used to support cultural diversity, in particular the multilingual nature of European society, by making language-learning easier and more accessible reducing the cost of translation and creating storehouses of cultural and linguistic materials, and so on. In particular, the accessibility of such materials and opportunities should be increased. However, in the short term, the English language is becoming more and more important in the international exchanges of the IS and tends even to become the dominant language used in some fields. A good example here is the Internet.
- ICT-based services can help to protect and reinvigorate local communities by reducing the cost and improving the quality of services which can be delivered in and by the community. In essence, distance can be shrunk so that people can live globally in their own communities.
- There is a proliferation of information transactions taking place because of the falling costs of production and, with the Internet, rapid and easy distribution of information. At the same time, there is concentration of ownership and resources taking place in the media. Clearly, therefore, there are major challenges in this area, given its importance as an economic sector in its own right, its centrality in cultural and political life, the problems of control over quality and reliability of information, and the need to protect the producers of information.

9. Democracy

There are great opportunities for an enrichment of democracy in the IS, through the development of horizontal exchanges of views and information between citizens, through greater information access, through more open and accessible government. However, the information explosion currently taking place, in terms not just of its quantity, but also in the variety of sources, formats, complexity and rapidity, could overwhelm people leading to an ephemeral and information overloaded political system. We, therefore, regard a coherent discussion of the challenges of the IS to democracy as an essential component of the debate in this area and provide some starting points for this debate below.

- There are great new opportunities for public expression and experimentation in the IS, especially in areas which are low cost and accessible to grassroots groups. More democratic and accessible government in particular is a major opportunity presented by these new developments.
- The vitality of political debate could, likewise, be reinvigorated through more use of direct democracy. To the extent that these systems can encourage horizontal communications and participation they should be encouraged, but the dangers of an ephemeral political culture, fuelled by the urgent immediacy of media, must be guarded against. There must also be space for sober reflection and careful decision making.
- The introduction of new modes of public opinion collection and information dispersal, perhaps through on-line systems, could disadvantage some groups in society. So, careful monitoring of all innovations in the democratic machinery is necessary in order to protect the democratic basis of society.
- Structural changes to the world economy (see amongst others Freeman and Hagedoorn, 1994; Freeman, 1996), the breakdown of traditional class structures and the growing internationalisation of society and the economy, are fuelling changes in the traditional political landscape. ICTs reinforce this change, especially through the promotion of industrial and social changes, and through the growing importance and immediacy of the international arena of politics.
- Protection of the democratic principle also comes from having a more articulate and educated population. Young people in particular need to be educated so as to be able to differentiate important messages from the noise, the real from the false, and the long-term from the ephemeral.
- Protection against the invasive potential of the new ICTs is also important both through direct legislative action and by building up, particularly, the international component of data protection legislation.

10. Conclusions

Fifteen years have passed since this text was written. Many of the ideas, expectations as well as concerns about what was then called the Information Society are still with us today. Those ideas, expectations and concerns were written very much from the perspective of a highly developed region such as the European Union. Over the last decade those expectations and concerns have undoubtedly continued to dominate the policy debate in the OECD world: work, employment, leisure, services, education, training, communication, etc, have all been influenced by the accelerating rate and speed of new, digital information and communication technologies. The implications for the emerging world, and in particular Africa, were not addressed. Yet the emergence of a virtual world side by side the real world offers probably one of the most fundamental development challenges in need of analysis.

As argued above, the information society has now become a knowledge based learning society. It will be to the new generation of innovation scholars from emerging countries and in particular Africa, to pick up from the insights presented above those which appear in particular relevant to development.

Notes

1 For further elaborations on this point see in particular Freeman's (1992) *Economics of Hope*.

2 This section was very much inspired by the Freeman and Soete (1994) Pinter book on *Work for All or Mass Unemployment*.

3 In one of his masterpieces with Francisco Louçã, Freeman elaborates on some of those issues; see Freeman and Louçã (2001) within the context though of the historical implications of the ICT revolution.

4 It is probably here that, after 1995, our analysis of the IS went in different directions. For a more radical, more digital economy view see amongst others Soete (2000), Soete and Weehuizen (2004) and Soete and Ter Weel (2005).

References

Freeman, C. (1989), 'Technical Change and Productivity', in *Finance and Development*, 26(3): 46-49.

Freeman, C. (1991), 'Technology, Progress and the Quality of Life', in *Science and Public Policy*, 18(6): 407-418.

Freeman, C. (1992), *Economics of Hope*, London: Frances Pinter Publishers.

Freeman, C. (1994), 'Technological Revolutions and Catching-up: ICT and the NICS', in Fagerberg, J., Verspagen, B., and Von Tunzelmann, N. (eds.), *The Dynamics of Technology, Trade and Growth*, Aldershot, UK: Edward Elgar.

Freeman, C. (1995a), 'Technical Change and Unemployment: The links between macroeconomic policy and innovation policy, innovation in a new context', in

STI Review (OECD, Paris), (15):.50-74.

Freeman, C. (1995b), 'Rise of East Asian Economies and the Computerisation of the World Economy', in *Journal of Economics and Business Administration*, 70(6): 1-21.

Freeman, C. (1996), 'Catching up and Falling Behind: The case of Asia and Latin America', in De La Mothe, J. and Paquet, G. (eds.), *EvolutionaryEconomics and the New International Political Economy*, London: Pinter Publishers.

Freeman, C. (and Foray, D.) (eds.) (1993), *Technology and the Wealth of Nations*, London: Frances Pinter Publishers.

Freeman, C. and Hagedoorn, J. (1994), 'Catching up or Falling Behind: Patterns in international interfirm technology partnering', in *World Development*, 22(5): 771-780.

Freeman, C. and Louçã, F. (2001), *As Times Goes By: From the Industrial Revolutions to the Information Revolution*, Oxford: Oxford University Press.

Freeman, C. and Perez, C. (1988), 'Structural Crises of Adjustment: Business Cycles and Investment Behaviour', in Dosi, G., Freeman, F, Nelson, R., Silverberg, G. and Soete, L. (eds.), *Technical Change and Economic Theory*, London: Pinter. pp. 38-66.

Freeman, C. and Soete, L. (1990), 'Fast Structural Change and Slow Productivity Growth: Some paradoxes in the economics of information technology', in *Structural Change and Economic Dynamics*, 1(2): 225-242.

Freeman, C. and Soete, L. (1994), *Work for All or Mass Unemployment*, London and New York: Pinter Publishers.

Freeman, C. and Soete, L. (eds.) (1987), *Technical Change and Full Employmnet*, Oxford: Basil Blackwell.

Freeman, C., Clark, J. and Soete, L. (1982), *Unemployment and Technical Innovation: A study of long waves and economic development*, London: Pinter.

Freeman, C., Soete, L. and Efendioglu, U. (1995), 'Diffusion and the Employment Effects of Information and Communication Technology,'in *International Labour Review*, 134(4-5): 587-603.

Soete, L. (2000), 'Towards the Digital Economy: Scenarios for Business', in *Telematics and Informatics*, 17: 199-212.

Soete, L. and Weehuizen, R. (2004), 'The Economics of E-government: A bird's eye view', in *International Journal of Communications Law and Policy*, 8.

Soete, L. and Weel, B. T. (eds.) (2005), *The Economics of the Digital Society*, Aldershot, UK: Edward Elgar.

Note from Luc Soete:

This text was written in 1995 by the late Christopher Freeman and myself as members of a High Level Expert Group on the Social and Societal Aspects of the Information Society, for the European Commission's Directorate General on Employment, Industrial Relations and Social Affairs. The text which became published as the Executive Summary of the Interim Report and was never published in an academic journal, illustrates neatly the broad spectrum of Christopher Freeman's vision on the way scientific and technological change, limited here to new information and communication technologies which were on the verge of impacting the global economy, would affect society. It is also not just an inspiring text in content but also in style. References to relevant publications to some of Freeman's earlier and later writings have been added.

Acknowledgement:

This chapter is based on the First reflections of a High Level Group of Experts entitled "Building the European Information Society for us All", published in January 1996, and prepared by Chris Freeman and Luc Soete. Comments at the time received from the other experts and in particular Gerhard Bosch and Manuel Castells are greatly acknowledged.

Chapter 5
How Much Is Free? Quantifying Open Source Software Development in South Africa

Michael Gastrow & Saahier Parker

1. Introduction

Open Source Software (OSS) is in essence a form of software where the source code of the software is available to users, and it is therefore possible for users to modify the software to suit their needs. In many cases this software is also free, or has weaker intellectual property rights (IPRs) attached in comparison to traditional proprietary software (for a more detailed definition see Krogh and von Hippel, 2003).

OSS has already had an enormous impact on global economies and cultures. The internet is based upon open source software (TCP/IP and HTML), which has contributed to its rapid growth and dynamic development. Developing countries, most conspicuously China, now widely use open source software. In the case of China this is manifested in the widespread adoption of the Linux operating system. This signals a slow revolution in the economies of software development and the related intellectual property rights. However, in most developing economies, including South Africa and other African countries, proprietary forms of software predominate. The intellectual property rights are almost entirely held by foreign enterprises, which in effect positions developing country firms as franchisees.

In the case of emerging economies, issues of development are inextricably linked with the manner in which ICTs, including OSS, are managed. For example, by expanding the scope for local innovation, an open source development environment allows local enterprises both to germinate and to move up the international ICT knowledge value chain. The use of OSS is believed to have many further benefits, with the most obvious being lower overall costs. Thus, while OSS is far from a panacea for the developing world's ICT challenges, it arguably has a major role to play.

Policies that aim to leverage OSS for development have been initiated in a number of developing economies, including South Africa. However, in the case of South Africa these policies have been developed in the context of a paucity of relevant data. There are currently no available data describing the extent of OSS uptake in South Africa, whether measured by usage or by expenditure on development. There are no national data with respect to government usage of OSS, the spread of Linux as an operating system, or the human resources employed in OSS systems and development. This means that government policies exist in a measurement vacuum, and since there are no base-line measurements, progress cannot be monitored. Also, a lack of local data makes international comparative research difficult, and it is thus a challenge to situate South Africa in an international context.

This chapter therefore offers a quantitative overview of OSS development in South Africa, drawing on recently released primary data from the South African National R&D Survey. This is an annual national survey of all R&D performing organizations in the country. In this instance, software development is included in the 'development' component of R&D, and is therefore measured by this survey. Moreover, since OSS is a national priority area, this form of software development has recently received a dedicated question in the survey. These quantitative findings are supplemented by interview notes from meetings with product development executives from four of South Africa's largest software development firms. Together these findings specify what we currently know with respect to OSS data in South Africa, and indicate what we ought to find out in terms of directions for further research.

This chapter is organized into six sections: Section two provides an overview of OSS in emerging economies; Sections three and four discuss OSS policy in South Africa and the issues related to quantifying OSS development respectively; Section five discusses national R&D data for South Africa and analyses R&D effort towards OSS development; and the final section draws some conclusions and makes some recommendations.

2. Open Source Software in Emerging Economies

Grandstrand (2000) describes a world economy that is moving from being industrial to knowledge-based, resulting in the emergence of 'intellectual capitalism'. In this context Information and Communication Technologies (ICTs) play a critical role, as they lower transaction costs associated with information flows and expand the reach of these information flows. ICTs have a particularly important role in the development of emerging economies: information technology supports development, both in the form of increased economic growth and in terms of human welfare (May, 2006). ICT is therefore a necessary component of development strategies for Africa. One of the key debates in the arena of ICTs in development concerns the role of OSS. This debate may be framed as a contest between investing in the current hegemony of proprietary software to drive ICTs, and the increasing recognition that OSS could play a larger role in development. For example, May (2006) reports that in 2006 approximately US$24 billion was transferred from African countries to non-African software companies, mainly based in the USA. As the ICT sector in Africa grows, and as the network predominance of proprietary software continues to grow, it is likely that the scale of such transfers will escalate.

Ghosh *et al.* (2002) identified four key benefits from the adoption of OSS by public institutions. These may be summarized as: (i) lower cost; (ii) independence from proprietary technology; (iii) availability of efficient and low-cost software; and (iv) capacity to develop custom applications and to redistribute the improved products. A move to OSS thus reduces financial outflows to foreign firms, and even in those cases where the short-term costs of OSS development are higher than proprietary software fees, users of OSS can avoid being locked in to buying future upgrades as well as the management costs of licenses, which reduces costs over the long term (National Advisory Council on Innovation, 2004).

Moreover, the funding required for local development circulates within the local economy, raising its multiplier effect and building human capital in the process. This is one of the key properties of OSS, in that enterprises whose core business is OSS will draw their primary revenue from downstream activities such as distribution and service, rather than income from legally protected intellectual property rights. Thus 'the essence of open source software is not the software. It is the process by which software is created' (Weber, 2004: 56).

In the public debate, proponents of OSS argue that it also supports entrepreneurship, business development, innovation, collaboration, local content and encourages a market for the development of software and media reflecting indigenous knowledge, culture and languages. Many governments, including those of developing economies, have instituted policies supporting OSS (Dravis, 2002). These countries include China, Japan, South Korea, Thailand, Argentina, Brazil, Peru, Germany, France, and the UK. It is widely recommended that 'developing economies and their donor partners should review policies for procurement of computer software, to ensure that options of low cost and/or open source software products are properly considered and their costs and benefits carefully evaluated' (Barton *et al.*, 2002: 105). South Africa's National Advisory Council concludes that OSS should be the 'default exploitation route for government R&D software' (Peeling and Stachell, 2001, quoted in NACI, 2004: 18).

Harnessing the benefits of OSS for development arguably requires that emerging economies must go beyond the use of Linux and other OSS software. Policy needs to be more comprehensive than simply encouraging the use of Linux in the public sector, as has been the case in countries such as Brazil and Thailand (Rossi, 2004). It needs to support the skills development that will allow for the local development of relevant applications. In order to attain such benefits, policies need to be based on a thorough understanding of OSS. This includes an understanding of the basic properties of OSS, as well as the incorporation of relevant data into analysis and policy-making processes (Camara and Fonseca, 2007). It is here that this chapter aims to make a contribution towards future efforts in measuring, analysing and informed policy making – firstly by outlining the scope of existing data, secondly by drawing tentative conclusions from these data, and thirdly by identifying directions for further research that would help policymakers develop appropriate interventions.

The need for such policy is underlined by the numerous impediments to the uptake of OSS in developing economies. Contextual challenges often include a low absorptive capacity and high failure rates (Ehikhamenor, 2003). Lobbying pressure from multinational proprietary software firms has a major impact on public procurement. There also remains the question of the impact of legacy software. Replacing such legacy software requires scarce financial and human resources, and also involves an element of risk. Thus, 'without a direct public policy for addressing these limitations, OSS will grow at the fringes of public and private companies and their core applications could remain based on proprietary software' (Camara and Fonseca, 2007: 123). This study speaks to the necessity of a direct public policy that is supported by sound quantitative and qualitative data.

3. Open Source Software and Policy in South Africa

With the changing of the South African political guard in 1994, the new government initiated an e-government programme that would enable "the continuous optimization of government service delivery, constituency participation, and governance by transforming internal and external relationships through technology, the Internet and new media" (Shilubane, 2001:40). In doing this the national government sought to use information technology as an enabling factor to improve service delivery by automating public access to government documents and services via multi-purpose web portals, personalized services and more efficient delivery of documentation and application processes. Though a number of key stakeholders were consulted in the establishment of the E-Government programme, including the National Advisory Council on Innovation, the State Information Technology Agency, firms and associations from the private sector, and internal government users, the program remains a national government initiative, administered through provincial government structures in all nine provinces.

In attempting to reduce the cost of these programmes, plans were made to save on dollar-based software licensing costs, where possible using local software products. This would also stimulate the local software and ICT industry. The Government Information Officers Council therefore recommended the adoption of Open Source Software as an alternative to proprietary software.

This recommendation galvanized a policy-making process that started with stakeholder consultation and the 2002 publication by the National Advisory Council on Innovation (NACI) of *Open Software and Open Standards in South Africa: A Critical Issue for Addressing the Digital Divide*. This submission to cabinet made Open Source Software a nominal non-negotiable base for the government ICT environment. This document informed South Africa's current Open Source Software policy that in principle is pushing toward adopting an open source operating architecture.

In 2006 a revised policy document was presented by NACI. This continued policy support for OSS in government, but was more prescriptive, highlighting themes, namely *Implementation, Migration, Development, Open Format Licensing, Promotion of The Wider Use of Open Source Software*. Each of these themes defined a particular phase in an implementation plan regarding open source software (see Table 1).

Table 1: South African OSS policy

Policy Statement Theme	Policy Wording
1. Implementation	The South African Government will implement FOSS unless proprietary software is demonstrated to be significantly superior. Whenever the advantages of FOSS and proprietary software are comparable FOSS will be implemented when choosing software solution for a new project. Whenever FOSS is not implemented, then reasons must be provided in order to justify the implementation of proprietary software.
2. Migration	The South African Government will migrate current proprietary software to FOSS whenever comparable software exists.
3. Development	All new software developed for or by the South African Government will be based on open standards, adherent to FOSS principles, and licensed using a FOSS license where possible.
4. Open Content format / licensing	The South African Government will ensure that all Government content and content developed using Government resources is made Open Content, unless analysis on specific content shows that proprietary licensing or confidentiality is substantially beneficial.
5. Promote the wider use of OSS	The South African Government will encourage the use of Open Content and Open Standards within South Africa.

Source: Compiled by authors

What have been the outcomes of these policies? This is a difficult question to answer without corresponding data: while currently available data provide a tentative measurement of OSS development activity, there exists no direct data describing institutional achievement with respect to policy. Gauging the success of these programmes would require monitoring and evaluation (M&E) research. This would require baseline data with respect to government and private sector usage of OSS platforms and applications, the links between these data and government policy initiatives, and qualitative data regarding the impact of these policies. Further to these baseline measurements would be longitudinal data reflecting the impact of these policies over time.

Despite a lack of monitoring data, a scan of institutional progress is a starting point that may provide some informative examples. In 2003 the Council for Scientific and Industrial Research (CSIR) launched the Open Source Centre within Icomtek, which is the information and communication technology hub at the CSIR. This centre had a number of projects assigned under its umbrella, ranging from software development projects to information and awareness dissemination of the benefits of open source among user communities. In May 2005 the Mareka Institute was established, also under the Council for Scientific and Industrial Research (CSIR), with a mandate to support awareness, adoption and development of Free and Open Source Software in research and development as well as implementation initiatives. The State Information Technology Agency (SITA) had at this stage already assisted some government departments with a limited conversion to OSS, but data that might quantify this conversion is limited or unavailable. It is thus clear that a more comprehensive institutional scan would be of benefit for OSS measurement and policy-making.

While government may be taking some action, it is the non-profit sector that has taken the lead in terms of OSS advocacy in South Africa. FOSSFA (Free Software and Open Source Foundation for Africa) was established in Geneva in 2003 in response to the UN and other world bodies calling for more input into African ICT and Open Source development and infrastructure. As its primary mission FOSSFA is mandated to promote the use of OSS in African development projects and co ordinate Africa's open source efforts to best utilize available resources and develop new centres of OSS excellence across the continent. In 2004 more than 250 delegates from around the world were gathered at the first FOSSFA-sponsored conference in Cape Town on OSS development for development in Africa, providing some momentum to the OSS movement in Africa.

The Shuttleworth Foundation has been arguably the most significant advocate of OSS in South Africa. Go-open Source, a two-year advocacy drive, waslaunched in 2004 to promote the use of open source software and provide the public with a resource for information and adoption of OSS technology. The program established a number of advocacy projects that ran until 2006. Tuxlab, another Shuttleworth Foundation project, was initiated in 2002 as a model for setting up computer facilities in South African schools using open source principles and software. However anecdotal evidence suggests that the government has hampered the roll-out of this project by supporting only Windows-based educational software. Other Shuttleworth programmes include the Linux Professional Institute and Hip2b^2, both of which have had a significant (if difficult to measure) impact on OSS in South Africa. This suggests another area of future research that might be of benefit to OSS policy-makers – an examination of the interaction between non-profit OSS projects and government institutions.

4. Quantifying Open Source Software Development

Measuring open source software development is challenging. The most common means of measuring the spread of open source software is by directly or indirectly measuring the level of market penetration of OSS. However, such an indicator is not sufficient for measuring the development of this software – a high level of market penetration may be drawing on OSS developed elsewhere. Other measures must be used to measure OSS development - for example expenditure levels. But such data are rare, and their collection and interpretation face significant challenges.

One such challenge is that a large proportion of OSS development takes place outside the formal working environment, and thus easily escapes measurement. A survey of OSS developers in Brazil found that approximately 40 per cent of developers had a paid job to develop OSS (Stefanuto and Salles-Filho, 2005, quoted in Camara and Fonseca, 2007). A similar proportion was found in the responses to a European survey by Ghosh *et al.* (2002). In both cases 60 per cent of OSS software developers were performing their work outside of the formal business environment, the so-called 'weekend' and 'basement' developers. This finding points to the fact that much of the work within the open source development area is outside of formal industry and therefore may not be effectively tracked and counted in surveys that seek to measure such industries, particularly within the context of developing countries. Therefore, any data drawn from the South African National R&D Survey may represent a lower boundary, and possibly only reflect 40 per cent of actual OSS development activity.

A third difficulty is that OSS development falls on the margin of what can be defined as 'software development'. The OECD Frascati methodology (OECD, 2002) describes software development as activity which produces novel software or adaptations leading to novel functionality in existing software. Interviews with executives from four of South Africa's largest software development firms highlighted that this resulted in difficulty in measurement, as much of their development activity would fall on the border of this definition. For example, customization of existing software may introduce new functionality, but classifying this as 'development' requires a careful inspection of the novelty of this functionality and whether this falls within the Frascati manual's definition thereof.

Despite these challenges, there are some available data to describe OSS in South Africa. These data are captured in South Africa's official national annual Survey of Research and Experimental Development Inputs performed by the Human Science Research Council's Centre for Science, Technology and Innovation Indicators (CeSTII), on behalf of the Department of Science and Technology. The survey is carried out according to the standard international Frascati methodology, and the resultant data form part of official statistics for the National Statistics System. Specialised questionnaires are directed at government, science councils, higher education institutions and non-profit organisations (NPOs). The business sector survey employs a purposive methodology, constantly expanding its database of firms that could potentially be involved in R&D in South Africa. Completion of the R&D survey, including basic economic data and extensive R&D data, is a statutory requirement for all organisations performing research or experimental development, including software development.

The R&D Survey is geared towards measuring national priority areas in science and technology. To this end the 2005/06 and 2006/7 R&D surveys requested that respondents estimate the percentage of R&D expenditure allocated to the development of open source software. The firms that answered positively to this question form the sample for this study.

R&D surveys are inherently imperfect, as they must rely on the co-operation of respondents. While the survey cannot claim to capture all the open source software development taking place within the country, it does include the major players, and is the most complete available data-set with respect to national open source software development. This includes data on expenditure, geographical location, sector, and collaboration.

These data must be seen as a lower bound, and the sample as representative of only those organizations that reported their activity to the national survey. The Science Council and Government sectors have relatively complete coverage, as all the Science Councils and all government departments are required to respond to the survey. Surveying is particularly difficult in the business sector, where firms do not have a strong incentive to disclose sensitive information. This is particularly applicable to open source software development, where international evidence suggests that a large proportion takes place in firms that may be too small to be included in the survey methodology. Individuals or micro enterprises are likely not to be covered by the purposive Frascati methodology.

With these limitations in mind, one can use these data to tentatively describe parameters of the scale, scope and structure of South African open source software development.

In order to provide a qualitative context for this data, the authors undertook interviews with product development executives and management teams from four of the largest software development firms in South Africa. These interviews were part of the extended fieldwork contributing to the

National R&D Survey. While confidentiality agreements preclude firm-level details, there were some general findings, which are valuable in setting a context for the R&D Survey data.

5. National R&D Survey Data

The interview findings thus set a context for the quantitative findings. A common point of discussion in all the interviews was that a national ICT skills shortage is having a major detrimental effect on firms developing OSS. The main challenge is attracting and retaining skilled staff. Even qualified staff often have a low level of skills. The root cause of this was identified as a poor quality of mathematics and science skills among those who are qualified. It was noted that because of an emphasis on certification, a South African qualification often does not guarantee applicable skills. Several examples were provided of work being outsourced to India due to a lack of local skills. An extreme shortage of previously disadvantaged staff also reportedly weakens these firms' Black Economic Empowerment (BEE) scorecards, which makes it more difficult to access government contracts and other forms of support. There were complaints about bottlenecks at the ICT Sector Education and Training Authority (SETA), including indefinite delays resulting in an inability to access funding for skills development. The skills shortage is particularly harmful in light of the highly competitive globalised market. A major cause of the underlying skills shortage was seen to be a lack of co-ordination between the Department of Science and Technology (DST), Department of Trade and Industry (DTI), and Department of Education (DOE), resulting in a lack of common understanding of the problem or the solution thereto. High levels of staff turnover in government reportedly exacerbates the problem.

There was a common response that the policy environment in South Africa lacks knowledge and fails to appreciate the challenges facing local IT firms that strive to be globally competitive. Also, it was felt that the contribution of ICTs to the economy is undervalued in policy circles. As an economic growth enabler it was felt that ICT could make a greater contribution to the South African economy if given more direct policy support (and the implementation of this policy).

The only positive comments with respect to policy were that some of the respondents were making use of the recently instituted R&D Tax allowance, which supports R&D and thus OSS R&D. This may indicate a way forward. One feature of the R&D tax incentive as a supportive measure is that it places the initiative for participation and benefit in the hands of firms. This means that the instrument is less likely to be rendered ineffective by weak implementation by government.

Firm-level interviews highlighted a frustration with government, and available data suggest that government has a minimal level of direct involvement in OSS development. As indicated in Table 2, in 2006/7, out of a sampling frame of 858 R&D performing organizations, a total of 38 were engaged in the development of open source software. The majority of these were private sector firms (26) and higher education institutions (9). Only one government department reported any activity.

Table 2: Profile of sample frame and sample: 2005/6 and 2006/7

	2005/6		2006/7	
	No of organizations performing R&D in the sampling frame	**No of organizations performing OSS R&D in the sample**	**No of organizations performing R&D in the sampling frame**	**No of organizations performing OSS R&D in the sample**
Business	607	20	677	26
Science Councils	24	1	24	1
Higher Education	43	8	44	9
Not-For-Profit	28	1	30	1
Government	88	1	83	1
TOTAL	**790**	**31**	**858**	**38**

Source: Centre for Science, Technology and Innovation Indicators (CeSTII), Human Science Research Council, South Africa.

However, this is not to say that government policy has been unsuccessful, or that OSS development is not on an upward trajectory in South Africa. The sample of organisations involved in the development of OSS can be placed in the context of general software development data (see Table 3). The 2006/7 National R&D Survey indicates that 141 organisations reported software development activity, representing 16% of the total respondents to the survey. Of these, 73 were dedicated software development firms, in that 100% of their R&D budgets were dedicated to software development. The total national expenditure for software development R&D was R836 million, or 5.1% of total national R&D expenditure. This can be seen in comparison to a total expenditure of R632 million in 2005/6, or 4.5% of national R&D expenditure, indicating that software R&D as a proportion of all R&D has increased.

National OSS expenditure data are reflected in Tables 4 and 5, and again these data represent a lower bound within the Business sector. Approximately R192 million was spent on open source software development in 2006/7, up from R101 million in 2005/6, or a nominal increase of 89%. Whether this increase is part of a sustainable trend or a short-term fluctuation will be determined by the measurements of future national R&D surveys.

Of the 2006/7 total, approximately R118 million (62%) was spent by private sector firms. Government departments accounted for only R4 000, or less than 1%. Overall, the proportion of national software development that was open source was approximately 16.1% in 2005/6, and 23.1% in 2006/7. The two years of available data therefore indicate an encouraging change towards a higher intensity of open source in the arena of software development research in South Africa. Thus not only is software development a growing component of overall R&D in South Africa, but open source has grown rapidly within this sector, mostly driven by private sector R&D investment.

Whether these findings are a result of market forces or successful government policy can only be determined by further research. However, these existing data will be useful as they represent benchmarks that can be used to measure future success in bolstering national OSS development levels – particularly as an outcome indicating the success of national policy.

Table 3: National R&D, Software Development and OSS development profile

	2005/6	2006/7
Total expenditure on R&D	R 14,149,239,000	R 16,520,584,000
Total expenditure on software development	R 632,666,250	R 836,274,259
Software development as a % of all R&D	4.5%	5.1%
Expenditure on OSS software development	R 101,937,110	R 192,785,822
OSS as a % of all software development	16.1%	23.1%

Source: National R&D Survey, CeSTII; *Note:* 1US$ = R8,00.

Table 4: National Expenditure on Open Source Software Development R & D

	2005/6 OSS R&D		2006/7 OSS R&D	
	Total R&D expenditure	%	**Total R&D expenditure**	%
Business	R 60,476,010	59%	R 118,857,770	62%
Science Councils	R 6,034,520	6%	R 27,509,571	14%
Higher Education	R 27,723,430	27%	R 41,441,391	21%
Not-For-Profit	R 7,700,600	8%	R 4,972,990	3%
Government	R 2,550	0%	R 4,100	0%
TOTAL	**R 101,937,110**	**100%**	**R192,785,822**	**100%**

Source: CeSTII

Table 5: 2005 / 2006 National R&D Survey: Collaboration profile of firms involvedin Open Source Software Development

	Collaboration Mode	**Number of firms (n=14)****
South African collaborators	Higher Education Institutions	5
	Science Councils	6
	Government Research Institutes	1
	Members of own company/ Affiliated Companies	6
	Other Companies (including Specialist consultants)	6
	Not-for-profit organizations	2
International collaborators	Higher Education Institutions	2
	Science Councils	2
	Government Research Institutes	1
	Members of own company/ Affiliated Companies	2
	Other Companies (including Specialist consultants)	4
	Not-for-profit organizations	1

Source: CeSTII

The Business Sector survey instrument in the National R&D Survey includes some specialized questions directed at private firms. These include questions about collaboration, which allows for the collaborative propensities of these firms to be represented in Table 6. Contrary to expectations, only 14 of the 26 OSS R&D performing firms reported to be undertaking collaborative activity. However, this may be due to incomplete questionnaires being received from some firms. These data suggest that OSS development in South Africa, as elsewhere, is highly collaborative: those firms that did collaborate had an average of 2.7 different modes of collaboration, which indicated that they are part of collaboration networks. The most common collaborative partners were reported to be local associates or partner firms, higher education institutions and science councils. The most common international collaborators were other firms, indicating that some of South Africa's major OSS developers are part of international innovation networks.

The National R&D Survey also includes data describing the geographical location of R&D activity. In line with national trends in R&D expenditure, the main loci of activity were reported in Gauteng and the Western Cape. Thus OSS development in South Africa is not only collaborative but geographically concentrated in the private sector. In essence the core clusters of OSS development in South Africa are collaborative private sector firms operating in Gauteng and the Western Cape.

Table 6: Geographical location of organisations performing Open Source SoftwareDevelopment R & D

2006-2007	Science Councils	Not-for-profit	Higher education	Government	Business	TOTAL
Eastern Cape			1			1
Free state			1			1
Gauteng	1		4		17	22
KwaZulu-Natal			1		1	2
Limpopo						
Mpumalanga						
Northern Cape				1		1
North West						
Western Cape		1	2		8	11
TOTALS	1	1	9	1	26	38

Source: CeSTII

6. Conclusions

This chapter has highlighted the limitations of our current knowledge about OSS in South Africa. While there exists limited data with respect to OSS development in South Africa (through the national R&D Survey), data describing market penetration of OSS platforms and applications are not available, nor are data describing the extent to which government OSS policies have been implemented. As a result, government policy is difficult to monitor or evaluate, making it difficult to determine whether these policies are on the right course.

This identifies some directions for future research that would benefit policy-makers. Firstly, primary research could be directed at government institutions to assess the extent to which OSS policy has actually been implemented – this could be through a more comprehensive institutional scan or through a more detailed follow-up of R&D Survey data. Firm-level surveys would be useful in measuring the market penetration of open source platforms and applications. Longitudinal data would be required for a sustained and reflexive monitoring and evaluation of government policies. Finally, the gathering of primary and secondary data describing the interaction between non-profit organisations involved in OSS advocacy and government institutions would be of benefit in understanding the dynamics between these key players.

However, some conclusions can be reached using currently available data. National R&D Data allow for some benchmarks to be set with which to monitor future performance. For example, national expenditure on open source software development increased by 89% between 2005/6 and 2006/7, from R101 million to R189 million. Approximately 23.1% of all software development reported in the 2006/7 survey was found to be open source, up from 16.1% in 2005/6. South Africa therefore appears to be progressing in terms of Open Source Software development capacity, although additional longitudinal data would be required to assess whether these changes are short-term fluctuations or part of longer term trends. Future national R&D Surveys will allow this comparison and allow for a measurement of the impact of national policies.

National R&D Survey data are indicative of the structure of OSS development in South Africa. OSS development is concentrated in private sector firms in Gauteng and the Western Cape. Organisations involved in OSS development are largely part of collaborative networks, both domestically and internationally. These factors suggest that more detailed firm-level research in these regions would greatly enhance our knowledge of OSS development, and may reveal means to harness the characteristics of these geographical and sectoral clusters, for example through industry organisations, cluster organisations, or other networking structures.

In the context of these data, firm-level interviews have been informative. Key findings were a dissatisfaction with government support for OSS, the identification of a skills shortage as a major impediment to OSS development, and the success of the national R&D Tax allowance in stimulating R&D. Interview-based research with a larger scale and scope may thus be highly informative, and could utilize existing R&D data to identify participants.

References

Barton J., Alexander D., Correa C., Mashelkar R., Samuels G., and Thomas S. (2002), *Integrating intellectual property rights and development policy*, London: UK Department for International Development, Commission on Intellectual Property Rights.

Camara G and Fonseca F. (2007), 'Information Policies and open source software in developing countries,' in *Journal of the American Society for Information Science and Technology*, 58(1): 121-132.

Dravis P. (2002), *A survey on open source software*, San Francisco, CA: The Dravis Group.

Ehikhamenor F.A. (2003), 'Information technology in Nigerian banks: The limits of expectations,' in *Information Technology for Development*, 10(1): 13–24.

Grandstrand O. (2000), 'The shift towards intellectual capitalism – the role of infocom technologies,' in *Research Policy*, 29(9):1061-1080.

Ghosh R.A., Glott R., Kreiger B., and Robles G. (2002)' *The free/libre and open source software survey and study—FLOSS*, Final report, Maastricht, The Netherlands: International Institute of Infonomics, University of Maastricht.

Krogh E., and Von Hippel E. (2003), 'Editorial: Special Issue on Open Source Software,' *in Research Policy,* 32(7): 1149-1157.

May C. (2006), 'Escaping the TRIPs' Trap: The Political Economy of Free and Open Source Software in Africa,' in *Political Studies*, 54(1): 123-146.

National Advisory Council on Innovation, Open Software Working Group (2004), *Free/Libre & Open Source Software and Open Standards in South Africa,* Stanford, California: Creative Commons.

Organization for Economic Cooperation and Development (2002), Proposed Standard Practice for Surveys on Research and Experimental Development Frascati Manual, Paris: OECD.

Peeling N. and Stachell J. (2001), 'Analysis of the impact of open source software,' Government IT Officers Council (Available at: http:/govtalk.gov.uk/library)

Rossi M.A. (2004), 'Decoding the free/open source (F/OSS) software puzzle: a survey of theoretical and empirical contributions,' in *Quaderni dell'Istituto diEconomia*, 424: 1–40.

Shilubane J. (2001), 'e-Governance: an overview,' in *Service Delivery Review*, South African Department of Public Service and Administration, Launch Edition:

40-44 (Available at: http://www.dpsa.gov.za/documents/service_delivery_review/Launch%20Edition/sdr_launch_ed_complete.pdf).

Stefanuto G.N. and Salles-Filho S. (2005), Impact of the free software and open source on the software industry in Brazil, Campinas, Brazil: UNICAMP.

Weber S. (2004), *The Success of Open Source*, Cambridge, MA: Harvard University Press.

Chapter 6
Research in African e-Local Governance:
Outcome Assessment Research Framework

Timothy Mwololo Waema, Winnie Mitullah, & Edith Adera

1. Introduction

The researchers who have worked on e-Governance have come from a variety of disciplines such as computer science, information systems, political science, and public administration. Using theories or frameworks from these disciplines, each comes with his/ her own perspective of what constitutes governance. While the body of knowledge is increasing, there has been no attempt to integrate this knowledge into a solid set of theories or frameworks that others can drawn upon; specifically, in evaluating outcomes or impacts of e-Governance. There is very little that has been produced in terms of a guiding theory or framework.

Instead, positivist statements about the benefits of e-Government and e-Governance dominate the literature. There are claims, for example, that e-Government can lead to the following outcomes: save costs while improving quality, response times and access to services (ADB, 2003); improve the efficiency and effectiveness of public administration (Pacific-Council, 2002); increase transparency in administration, reduce corruption and increase political participation (Seifert and Bonham, 2003); and make governments more competitive (OECD, 2003). Most of these are e-Governance outcomes.

The situation on the ground is not quite so rosy, however, as most e-Government projects or initiatives have not achieved the espoused outcomes. In 2002, for example, the Gartner Group reported that "more than 60 per cent of all e-Government initiatives either fail or fall short of expected outcomes."[1]

In addition, Saxena (2005) argues that in spite of the world wide diffusion of e-Government initiatives, getting the claimed benefits of e-Governance has not been easy for various technological as well as organizational reasons. Heeks (2003) notes too that e-Government projects often fail either totally or partially in achieving their objectives despite their initial successes. Positivist statements or conclusions on the benefits of e-Government and e-Governance initiatives, therefore, have largely been based on studies that have lacked a serious research framework for assessing the associated benefits or outcomes.

The research framework reported in this chapter was developed in an attempt to have a guiding model for assessing good governance outcomes of ICTs at the local government level in nine African countries under the Local Governance and ICTs Research Network for Africa (LOG-IN Africa)[2].

In addition, the study explores the following: a review of the key definitions used in the study; a description of results-based management by concentrating on the results chain and results-based outcome monitoring and evaluation; focus on the research framework for evaluating the outcomes of e-Governance; illustration of the use of the developed framework with a case study; and finally conclusions and recommendations.

2. Definitions

Governance and Good Governance

In most economically developing countries, the concepts of governance and good governance have taken root as part of the conditions imposed by the so called "development partners" in development aid as well as being in most cases part of a wider public sector reform programme that is often externally driven through the World Bank and other donor institutions. Even then, the concepts take different definitions depending on the interests of the "donors."

Definitions of governance vary widely depending on the source (e.g. GDRC, 2004a; 2004b; 2004c). All definitions agree on one thing: governance is not synonymous with government. For purposes of this chapter, government focuses on activities and outputs by expending effort while governance focuses on processes and outcomes effects produced over a longer time perspective.

Like governance, "good governance" has been defined differently by various authors and institutions. Many definitions are therefore subjective. Given that our assessment framework is on "good governance", it is crucial to make explicit the appropriate and workable definition of this concept.

The World Bank highlights four elements that constitute good governance[3]:

(a) openness and predictability in policy making

(b) professionalism in bureaucracy

(c) accountability of government and participation of civil society

(d) adherence to the rule of law

In addition, the World Bank also provides a list of indicators to measure good governance. However, we consider the elements of good governance as defined by the World Bank as inadequate to take care of the issues we would be concerned about in e-Governance implementation. Firstly, openness is only restricted to policy making. We acknowledge that policy formulation and implementation are highly interrelated and complex and it is often difficult to separate them. However, in e-Governance implementation, we are also concerned about the transparency of the implementation processes and the openness with which key stakeholders have access to information and knowledge. Secondly, participation is only restricted to civil society. We see multiple stakeholders, including private sector and the public, as participants in e-Governance, whether at the policy or implementation levels. Further, it is doubtful whether professionalism in bureaucracy would necessarily ensure efficient utilization of resources while serving stakeholders, an important good governance consideration. Finally, the World Bank definition of good governance excludes concepts that we deem important, including responsiveness and equity.

We now review good governance as defined by UNDP. UNDP describes good governance as being:

> ...among other things participatory, transparent and accountable. It is also effective and equitable. And it promotes the rule of law fairly. Good governance ensures that the voices of the poorest and the most vulnerable are heard in decision-making over the allocation of development resources, and that political, social and economic priorities are based on broad consensus among the three stakeholders - the state, private sector and civil society.[4]

Here, attributes of the concept of good governance are identifiable: (i) adoption of a participatory approach; (ii) transparency and openness; (iii) accountability through assuming responsibilities for actions; (iv) effectiveness, equity and fairness; and (v) endorsement of the rule of law: openness of decision-making to the input of vulnerable social segments as well as the formulation of the national agenda through a consensus between state, private sector, and civil society, all of which serve as the stakeholders in governance.

Most of the definitions of good governance tend to focus on the quality as well as the characteristics of the processes of "governance" (e.g. Weiss, 2000). The definition by UNDP (2002) in which the characteristics of good governance, together with the interpreted meaning, are outlined in the Table 1 below.

Table 1: UNDP's Nine Underlying Characteristics of Good Governance and Their Meaning

Characteristics of Good Governance	UNDP's Definition	Interpreted Meaning
1. Participation	All men and women should have a voice in decision-making, either directly or through legitimate intermediate institutions that represent their interests. Such broad participation is built on freedom of association and speech, as well as capacities to participate constructively"	The act of involving people, regardless of gender, to voice their interests in the decision-making processes. This involvement can be direct or indirect, e.g. the public participating through institutions that articulate their interests
2. Rule of Law	Legal frameworks should be fair and enforced impartially, particularly the laws on human rights"	The act of justly and objectively putting laws in place
3. Transparency	Transparency is built on the free flow of information. Processes, institutions and information are directly accessible to those concerned with them, and enough information is provided to understand and monitor them"	The full release of information and providing stakeholders with free access to institutions, operations and information
4. Responsiveness	"Institutions and processes try to serve all stakeholders"	Institutions and operations answer the requests of stakeholders

5. Consensus Orientation	Good governance mediates differing interests to reach a broad consensus on what is in the best interest of the group and, where possible, on policies and procedures"	Arbitration of the clash of interests in order to establish agreements based on the optimal interests of stakeholders
6. Equity	All men and women have opportunities to improve or maintain their well-being"	The act of providing people, irrespective of gender or other factors of possible discrimination, with equal chances to foster the quality of their welfare
7. Effectiveness and Efficiency	Processes and institutions produce results that meet needs while making the best use of resources"	Achievement of optimal use of resources while serving stakeholders
8. Accountability	Decision-makers in government, the private sector and civil society organisations are accountable to the public, as well as to institutional stakeholders. This accountability differs depending on the organisation and whether the decision is internal or external to an organization"	Managers and decision-makers are held liable to the community
9. Strategic Vision	Leaders and the public have a broad and long-term perspective on good governance and human development, along with a sense of what is needed for such development. There is also an understanding of the historical, cultural and social complexities in which that perspective is grounded"	Managers and the public have a long-term view with regard to governance and are aware of contextual obstacles related to history, culture and society

The list of indicators of good governance often tends to change according to the organisation defining them. We acknowledge that defining the principles of good governance can often be difficult and controversial, but in this study we think that UNDP's definition of good governance contained in Table 1 would be a good starting point for two main reasons. One, the definition contains a comprehensive coverage of all key characteristics of good governance. Two, the characteristics of good governance have been defined at a fairly generic level, which makes it relatively easier to customize them to the governance context one is handling. For example, there is no mention of any particular stakeholder group in participation. We have thus adopted this definition to include all the relevant factors for evaluation of e-local governance in this research.

The UNDP's definition of "good governance" provides a helpful heuristic, even though it is not necessary to have all the elements of the UNDP's constructs of good governance for an e-Government application to be recognized. The UNDP's definition is wide enough and allows us to investigate different manifestations of good governance in ICT implementations. Indeed, the research teams for whom we developed this assessment framework focused on different elements of the nine constructs depending on the e-Government application they were assessing.

E-Government and E-Governance

E-Government means different things to different people. Ndou (2004:6) identifies three key applications of e-Government:

- "e-Administration" in which an administration or a government office uses ICT in order to interrelate its various departments and digitize its internal operations via "automation and computerization of administrative tasks".
- "e-Citizens and e-Services" which are e-Government applications that enable online access to government information and knowledge and "deliver automated services".
- "e-Society" which provides a platform that facilitates interactions between government actors and civil society.

In this study, we take Ndou's (2004) broad categorization of e-Government as a good starting point. We shall refine this definition in describing our research framework, however. Like definitions of e-Government, definitions of e-Governance vary greatly depending on who defines it and what are the contexts of its definition. For purposes of this study, we take e-Government to be concerned with the application of ICTs to automate specific aspects of government functions and administrations. E-Governance as a concept is taken to mean beyond e-Government and concerns itself with the impacts of e-Government on governance improvement, effectiveness, efficiency and overall or performance. Thus it not only concerns itself with the mere application of ICTs but also with how these ICTs are used to achieve governance outcomes.

3. Outcome Monitoring and Evaluation

According to UNDP (2002), outcome monitoring refers to:

> ...a continual and systematic process of collecting and analyzing data to measure the performance of UNDP interventions towards achievement of outcomes at country level. While the process of outcome monitoring is continual in the sense that it is not a time-bound activity, outcome monitoring must be periodic, so that change can be perceived. In other words, country offices will accumulate information on an ongoing basis regarding progress towards an outcome, and then will periodically compare the current situation against the baseline for outcome indicators and assess and analyze the situation (UNDP, 2002:20).

How should outcome monitoring be carried out? According to UNDP, to monitor outcomes, managers need to track the outputs and measure their contributions to outcomes by assessing the change from baseline conditions. Further, to conduct effective outcome monitoring, managers need "to establish baseline data, select outcome indicators of performance, and design mechanisms that include planned actions such as field visits, stakeholder meetings and systematic analysis or reports" (UNDP, 2002:20). Monitoring outputs, therefore, is an integral part of outcome monitoring.

Outcome evaluation refers to:

> ...an evaluation that covers a set of related projects, programmes and strategies intended to bring about a certain outcome. Such evaluations assess how and why outcomes are or are not being achieved in a given country context, and the role UNDP has played. They may also help to clarify underlying factors affecting the situation, highlight unintended consequences (positive and negative), recommend actions to improve performance in future programming, and generate lessons learned. These periodic and in-depth assessments use "before and after" monitoring data (UNDP, 2002:21).

Monitoring and evaluations of results are performed through performance indicators. Each of the result levels has associated indicators. These are qualitative or quantitative "pointers" that can be measured or described and that help determine progress towards achieving planned results. It is through the regular measurement or observation of these key outcome indicators that we can determine if outcomes are being achieved.

Kusek and Rist (2004) developed a ten-step approach to designing, building and sustaining a results-based monitoring and evaluation system:

(i) Conducting a readiness assessment

(ii) Agreeing on outcomes to monitor and evaluate

(iii) Selecting key indicators to monitor outcomes

(iv) Collecting baseline data on the indicators

(v) Selecting the targets for the outcome indicators

(vi) Monitoring

(vii) Evaluating

(viii) Reporting

(ix) Using the findings

(x) Sustaining the M&E system

In the proposed research framework, we shall take a subset of these steps for monitoring and evaluating good governance outcomes of e-Government projects. The purpose of monitoring and evaluating the outcomes in the case of LOG-IN Africa is to test the usefulness of the proposed research framework, learn some lessons of the effects of e-Government initiatives or projects on good governance and create knowledge that can be used to build a roadmap on e-local governance in Africa.

4. e-Governance Outcome Evaluation Research Framework

Kettani *et al.* (2005) created a methodological framework for assessing the impact of e-Government systems and governance. The framework does this by assessing the impact of e-Government on good governance outcomes, among other major outcome categories.[5] The researchers used UNDP definitions of the attributes of good governance but customized the definitions to fit the e-Government project that they were studying: a system for providing citizen-related services. They then developed measurable set of indicators for the identified project outcomes. The idea was to assess how project activities and outputs generate project-defined good governance outcomes using defined indicators and compare these with values of indicators before the e-Government system was introduced (or what is commonly referred to as reference or baseline status).

The work of Kettani *et al.* (2005) was our starting point for developing the e-Governance outcome evaluation framework reported here[6] because, like Kettani *et al.*, we took UNDP's definition of good governance as it is inclusive and contains comprehensive coverage of all key good governance characteristics. We also agreed with the researchers on the need to create "working" or contextualized definitions of good governance. Our proposed research framework has two notable differences from the work of Kettani *et al.*

The first point of departure from the work of the researchers under reference is that we have taken into account the various dimensions of e-Government. In our framework, we take the three applications Ndou (2004) defines (as outlined earlier) as reflective of all key dimensions of e-Government. For purposes of our framework, we re-define these components thus:

> **E-Administration**: the computerization of internal government business processes, including business process re-engineering and the development of the required ICT infrastructure;
>
> **E-Services**: the electronic delivery of government information, knowledge and services to citizens, businesses and other government units through wide area network infrastructure and Internet connectivity as may be required; and
>
> **E-Society**: the electronic interactions between government actors and citizens, businesses and other government actors. This interaction can be through a number of communication facilities, including mobile phones, cyber cafes and public access facilities.

The key reason we have added this component to our framework is our conviction that good governance outcome indicators will vary depending on the phase an e-Government project is in. For example, the good governance indicators for participation in the e-administration phase of an e-Government project are different from the indicators in the e-services phase of the project. This will become more obvious when we later provide indicators for the various components or phases of e-Government.

The second point of departure from that work is the use of results-based outcome monitoring and evaluation approach to create, monitor and evaluate the good governance outcomes. More specifically, we have borrowed from Kusek and Rist's (2004) monitoring and evaluation steps outlined earlier. We have proposed a *five-step iterative process* as a methodological approach to evaluate good governance outcomes. This approach is outlined below:

1. **Assessment of context**: Conduct an assessment of the context of the e-Government project to gain a deeper understanding of the context of the implementation for the e-Government intervention being studied. The context may include the e-Government/e-Governance policy environment, business and ICT environment, political commitment, ICT infrastructure and connectivity, human resource skills and budget allocation. This context becomes key in explaining why or why not certain outcomes were or were not realized.

2. **Working definitions of 'good governance':** Create contextualized or "working" definitions of each of UNDP's broad constructs of 'good governance' that the research will focus on. These localized definitions would inevitably be linked to the national government's and/or local government's concepts of good governance as well as the specific context of the e-Government project.

3. **Outcome indicators**: Translate the working definitions of good governance, into identifiable and measurable (or at least observable) outcome indicators. This translation is carried out for each of the three dimensions of e-Government. Outcome indicators can be further specified or associated with identifiable and measurable output indicators which are meaningful in the specific context and directly linked to the e-Government project based intervention.

4. **Baseline data on indicators:** Collect baseline data on the chosen indicators to establish the status of the various outcome indicators before the implementation of the e-Government intervention. This status sets the reference against which future changes can be assessed.

5. **Evaluate the outcomes**: This is the assessment of the status of the chosen indicators to establish the effects of the completed e-Government intervention on good governance. In general, outcome measurement is a challenge (Mayne, 2007) and indeed this is where the greatest challenge lies. In particular, determining the extent to which an e-Government intervention can be associated with observed outcomes is quite a challenge, largely because of the attribution problem. Further, the long periods from the times outputs are produced to the time "allegedly" associated outcomes are realized creates an additional problem of attribution. In order to deal with these challenges, we propose that researchers focus on the degree to which an e-Government system under study has *contributed* to the status of the outcomes in question, and, at the same time, establish *otherfactors* that have contributed to the realization of the outcomes.

The above approach is necessary in order to have a sound methodological framework for monitoring and evaluating e-Governance outcomes.

Table 2: Proposed Research Framework

Good Governance Constructs	Working Definitions of Good Governance	Outcome Indicators		
		E-Administration	E-Services	E-Society
Participation				
Transparency				
Responsiveness				
Equity				
Effectiveness & Efficiency				
Accountability				
Rule Of Law				
Consensus Orientation				
Strategic Vision				

Table 2 presents a representation of the proposed conceptual framework, combining the constructs or characteristics of good governance, components of e-Government and the concept outcome indicators across the three aspects of e-Government for outcome monitoring and evaluation. It is to be noted that outcomes will vary from one project to another and therefore a tuning of the constructs of good governance to be more relevant to the project context. This implies the need for a working definition that is more functional, more precise and contextualized to the project under evaluation.

5. Illustration of the Framework

Generic Outcome Indicators

At a generic level, the LOG-IN researchers collectively developed a set of generic good governance outcome indicators. These are illustrated in Table 3. Each country project team was then supposed to translate these generic indicators into the relevant project-level outcome indicators for that country and in line with the methodological framework outlined above.

Table 3: Generic Good Governance Outcome Indicators

Good Governance Constructs	Working Good Governance Definitions	Good Governance Outcome Indicators		
		E-Administration	E-Services	E-Society
Participation	Stakeholders participate in local government decision-making processes and activities	• Extent of participation in decision-making of staff at different levels	• Extent of involvement of customers, politicians, private sector and civil society in LG e-Governance processes • Extent of availability of electronic delivery facilities empowering citizens to serve themselves without being dependent on employee-intermediaries and their "good will" to do their job	• Extent of improvement in interactions with central government, businesses, citizens and other key stakeholders • Degree of joint planning between local government and civil society • Extent of less privileged people, including illiterate and indigenous people, participating in LG decision making

Table 3: Generic Good Governance Outcome Indicators (continued from previous page)

Good Governance Constructs	Working Good Governance Definitions	Good Governance Outcome Indicators		
		E-Administration	E-Services	E-Society
Transparency	The full disclosure of information, presentation of information in an easy and accessible format, and free access to LG operations and information by stakeholders	• Degree of access and understanding of e-Governance policy and strategy by stakeholders • Extent to which LG business processes and information are clear and accessible to stakeholders • Degree of visibility and trackability of LG internal operations • Extent to which government institutions exchange and share information	• Extent of improvement of public access to pertinent information and services • Degree of ease for citizens to find/ access procedures to follow in order to request and receive services • Extent to which ordinary citizens can read and understand the language by which information and service are presented • Degree of ease for citizens to post comments, questions, and complaints regarding service delivery • Response rate to citizens' comments, questions, and complaints • Degree of availability of government official documents as electronic resources • Extent of availability of the public servant's profile	Not applicable

Table 3: Generic Good Governance Outcome Indicators (continued from previous page)

Good Governance Constructs	Working Good Governance Definitions	Good Governance Outcome Indicators		
		E-Administration	E-Services	E-Society
Equity	Services provision or access on equal basis irrespective of gender, disability, socio-economic status or other forms of possible discrimination	• Degree of equity in computerized service access by staff without discrimination • Degree of equity in staff ICT training without discrimination	• Degree of equity in service delivery irrespective gender, country's geographical localization and other forms of marginalization • % reduction in the frequency of "favouritism" incidents in serving customers • Extent of availability of "self-service-technology" (e.g. touch screen kiosks) to empower citizens by enabling them to serve themselves without being dependent on employee-intermediaries as a way to further fight favouritism incidents and endorse the principle of "first come first served"	Not applicable

Table 3: Generic Good Governance Outcome Indicators (continued from previous page)

Good Governance Constructs	Working Good Governance Definitions	Good Governance Outcome Indicators		
		E-Administration	E-Services	E-Society
Accountability	LG managers and employees are accountable to citizens and other key stakeholders	• Extent of accountability of LG managers and employees for their actions • Extent of availability of ways to monitor and trace LG personnel activities (i.e. know who does what and when)	• Extent to which LG staff can be identified with their actions in service delivery and be held responsible for them • Degree to which public servants and local governors account to the citizens	Not applicable

Table 3: Generic Good Governance Outcome Indicators (continued from previous page)

Good Governance Constructs	Working Good Governance Definitions	Good Governance Outcome Indicators		
		E-Administration	**E-Services**	**E-Society**
Consensus Orientation	Conflicts of interest of the various stakeholders are resolved	• % reduction in internal conflicts of interest • Extent of reduction in the possibility of alienation of staff that could be caused by minority versus majority of votes	• % reduction in conflicts of interest in service delivery	Not applicable
Strategic Vision	Managers and the public have a long-term view of governance but are aware of contextual obstacles and their solutions	• Degree of enhancement in long-term view of governance • Extent of existence of e-Governance strategy • Extent of alignment of e-Governance strategy to national socio-economic development priorities	• Degree of enhancement in long-term view of governance • Extent of existence of e-Governance strategy • Extent of alignment of e-Governance strategy to national socio-economic development priorities	Not applicable

As a further illustration of the use of the above framework, we have provided a brief case study of Kenya below. This case study focuses on an integrated financial management system implemented in two local governments.

6. Case Study of Kenya

National ICT Context

The Government of Kenya, recognized the importance of ICT in 2004, when it launched an ambitious three year (2004-2007) e-Government strategy. This strategy was designed to achieve a set of goals and objectives, namely: to efficiently deliver government information and services to the citizens; to promote productivity among public servants; to encourage participation of citizens in government; and to empower all Kenyans in line with the development priorities outlined in the 2003 – 2007 Economic Recovery Strategy for Wealth and Employment Creation.

The e-Government priorities include: instituting structural and operational reforms; review of the regulatory and legal framework; and development of a reliable and secure infrastructure. The priority activities and implementation framework, over the immediate, medium and long terms include: Government to Government (G2G); Government to Business (G2B); and Government to Citizen (G2C). In G2G, the strategy aims at expanding the information infrastructure, initiating the integration of internal government processes, and to increasing the efficiency and effectiveness of internal operations in the short term. In the medium term (by 2007), the plans include completion of the information infrastructure within the government up to the district level, developing and implementing web-enabled databases, and operationalizing other information systems under development, e.g. civil registration, road transport, revenue collection, etc.

The E-Government strategy further aims at establishing electronic forums for discussion and feedback in G2B and increased involvement of citizens in decision-making and public activities in G2C in the medium term (by 2007). Unfortunately, the E-Government strategy did not mention local authorities. According to the E-Government Permanent Secretary, local authorities were to be addressed through their parent ministry, the Ministry of Local Government. The strategy came to an end in 2007, and its achievements have been evaluated and a draft strategy for the next five years created. By the time of writing this chapter, the next e-Government strategy had not been officially sanctioned.

The Kenyan government worked on an ICT policy for several years (Waema, 2005). Indeed, the past years have witnessed significant policy activity in this arena with the finalization of an E-Government strategy, the issuance of a universal access report, the development of two ICT policy drafts, and most importantly, the acceptance of the Government of Kenya of the need for multi-stakeholder participation in the drafting process. The policy was discussed and approved by the Cabinet in January 2006 and an ICT policy document published in March 2006. Interestingly, the e-Government strategy had been created earlier.

The national ICT policy highlights the overall goal of E-Government, which is to make the Government more result oriented, efficient, and citizen centred. The E-Government strategy focused on redefining the relationship between Government and citizens, with the objective of empowering them through increased and better access to government services. The policy gives the broad objectives of E-Government as:

(a) Improve collaboration between Government agencies and enhance efficiency and effectiveness of resource utilization;

(b) Improve Kenya's competitiveness by providing timely information and delivery of Government services;

(c) Reduce transaction costs for the Government, citizens and the private sector through the provision of products and services electronically; and

(d) Provide a forum for citizens' participation in Government activities.

It is hoped that the new e-Government strategy to be soon launched will be in line with the national ICT policy. The current strategy had been developed ahead of the national policy, which was a demonstration of the disjointed approach to policy and strategy. It is also hoped that the new e-Government strategy will highlight the role of LAs in delivering services to citizens and effecting local social and economic development.

Local Governance Context

The Kenyan local governance system is composed of the Ministry of Local Government and the four tiers of LAs, namely: Cities, Municipalities, Towns and County Councils. These councils are corporate entities that are established under the Local Government Act Chapter 265, which is currently under review. In addition to the Act, the LAs draw their legal powers from the Constitution of Kenya, other Acts of Parliament, Ministerial Orders and By-Laws.

While the above legal bodies constitute Local Government in Kenya, the local governance framework in Kenya is broader than the LAs. It consists of provincial, district, location and sub-location administration with technical staff drawn from various ministries. Inherent in the gamut of Local Government and the local governance system are various public and private institutions, including civil society organizations. Kenya has no decentralization policy that rationalizes power sharing, responsibilities, and resources between the central government ministries, parastatals, District Development Committees, LAs and the civil society organisations. This has been a problem at operational local level, with most of the institutions and organizations lacking synergy.

LAs face a number of challenges in realizing their mandate. The challenges include: delivery of infrastructure and services; financial management; institutional and legal framework; human resource capacity; and managing rapid growth. These challenges have resulted in poor service provision and management with many analysts criticizing the LAs, and questioning their role in local development. It is this inefficiency that justified a re-examination of their role and the launching of the LGRP which influenced the LAs to embrace new forms of management including adoption of computerized financial management.

Since the beginning of 1990, when the Government of Kenya began implementing Structural Adjustment Programmes (SAPs), followed by civil service sector reforms in 1993, a number of reforms have been realized. In particular, there has been a deliberate attempt to decentralize governance by moving away from a centralized political system where citizens hardly make any contribution in governance of resources, to a decentralized political system where the inputs of citizens is considered critical to development (Rockeliff and Mitullah, 2003). This has witnessed the review of the Constitution, which stalled, the review of the Local Government Act Chapter 265, which has been on hold pending the completion of the review of the Constitution; review of various Acts of Parliament; the development of Strategic Plans by Government Ministries and parastatals; as well as the adoption of Result Based Management (RBM) by public institutions.

In 1996, a decentralization initiative under the Local Government Reform Programme was launched, it focused on strengthening LAs. The programme has three components: rationalizing central-local financial relations; improving LA financial management, including revenue mobilization; and strengthening citizen participation in planning and ownership of programmes. The reform programme has recognized the importance of LAs in enhancing economic governance, improving public service delivery, and increasing economic efficiency, accountability and transparency (ROK, 1999). The reforms have also included putting in place Fuel Levy Fund, Contribution in Lieu of Rates, user charges rationalization, Single Business Permits (SBP) and the IFMIS. These programs aim at restructuring the local public sector and more importantly, strengthening local level accountability mechanisms.

KLGRP began with financial reforms aimed at enhancing inter-governmental fiscal transfers, improving financial management, debt resolution, streamlining budgeting systems, service provision, and capacity building for LAs. A key instrument in this process was the enactment of the Local Authority Transfer Fund (LATF) in 1998. The LAs began accessing LATF during the financial year 1998/1999. The Act provides 5 per cent of national income tax to LAs, in line with population, resource base and financial performance. In order to access LATF, LAs are administratively required to develop a Local Authority Service Delivery Action Plan (LASDAP) using a participatory approach. Disbursement of LATF funds depends on strict conditions which if adhered to, would not only improve financial management but greatly enhance accountability and citizens' participation in the affairs of the councils.

Grants from central government include LATF, Road Maintenance Levy Fund (RMLF) and Contribution in Lieu of Rates (CILOR). Prior to setting up of LATF, there was limited sharing of resources with central government. Since the 1999/2000, financial year, LAs have been accessing LATF. The LATF is aimed at complementing the revenue of LAs, which are mainly two: (i) grants from the central government; and (ii) revenue from local sources. Local revenue includes: user charges and taxes from local sources (e.g. property rates); Single Business Permits (SBP); plot rent; market and bus park fees; water and agricultural cess; national reserve and park fees. Municipalities rely more on water and sewage fees, property rates and single business permits while county councils and town councils rely on single business permits, market fees, bus parks and cess revenue.

Implementation of IFMIS

The Integrated Financial Management Information System (IFMIS) is part of the Public Sector reforms focusing on improvement of the public sector, including local governance in Kenya. The Kenya Local Government Reform Programme (KLGRP) is a component of the public sector reform. One of its components is the IFMIS, which aims at improving financial management and revenue mobilization in local authorities. The two cases studies - Mavoko and Nyeri Municipal Councils were the first to benefit from IFMIS in 1999.

Since 1999, the implementation of IFMIS has been extended to six other LAs, namely: Karatina; Embu; Eldoret Municipal Councils; Kiambu; Wareng; and Kirinyaga County Councils. In these councils, IFMIS is used for all financial management activities, including the billing and collection of all local revenues, payroll, and all expenditure controls including issuance of vouchers, procurement, payment of statutory debts and project expenditure management. The financial administrative activities are linked through the budget monitoring system, generating a series of operational and management reports to assist in controlling, monitoring, and managing all financial activities within the LA. The IFMIS began as a national programme and was later broadened to include LAs and hence the concepts Local Authority Integrated Financial Operations Management System (LAIFOMS)[7]. The effect of this system, on good governance, is the subject of this study.

7. Methodology

Data for this case study was collected from both secondary and primary data. The secondary sources included a review of literature, including government and academic publications, and reports; while the primary information gathering included surveys, key informant interviews and Focus Group Discussions (FGDs). Primary data using survey method was gathered from a sample of staff from the two councils; consumers of the Council services in both councils; citizens and businesses within the jurisdictions of the two councils; and key informants drawn from both councils and central government officials.

In total, 36 council staff questionnaires and 104 consumer questionnaires were administered. Information gathered using these methods was triangulated with outputs from Key Informants interviews and FGDs with senior officers of the councils, councillors and consumers as separate groups.

Good Governance Outcome Indicators

The outcome indicators for the e-Governance assessment study that was carried out for the financial management application in Kenya are shown in Table 4. This study focused on five constructs of good governance[8] in two municipal councils in Kenya (Waema and Mitullah, 2007). The study was carried out by conducting an assessment of the e-Governance context, creating working definitions of the five good governance constructs chosen for this project, creating the outcome indicators pertinent to the project, and evaluating the e-Governance outcomes as per the indicators. Baseline data for the indicators were collected through historical reconstruction as this was a project that had been implemented. The study found that the above framework was useful in assessing good governance outcomes of e-Government projects. More specifically, it found that the integrated financial management system in the two municipal councils was fairly successful in realizing positive good governance outcomes in responsiveness, accountability, transparency and efficiency and effectiveness. Although the study found a significant increase in citizen participation following implementation of the system, focus group discussions established that this change could not be attributed to the system implementation.

8. Conclusions and Recommendations

This study has viewed good governance as the quality of governance. It has adopted UNDP's definitions of the various qualities of governance as a starting point in defining the concept of good governance, in the process identifying and defining three main components of e-Government: (i) the computerization of internal government business processes; (ii) the electronic delivery of government information, knowledge and services to citizens, businesses and other government units through wide area network infrastructure and Internet connectivity as may be required; and (iii) the electronic interactions between government actors and citizens, businesses and other government actors.

Further, the paper has conceptualized e-Governance as the relationship between the various components of e-Government and a well-defined concept of good governance to produce governance outcomes. With these definitions, it has presented results-based monitoring and evaluation approach as a good basis for investigating the effects of e-Government on good governance. At the same time, it has integrated the defined concepts and proposed a research framework for assessing the good governance outcomes of e-Government projects. We have illustrated the use of this framework with the developed generic good governance indicators at the pan-African level and the specific indicators created using the Kenyan case study.

We believe that the proposed research framework will help researchers to define a set of e-local good governance outcome indicators. This will be done through a process of aggregating, integrating and generalizing outcome indicators from the various national projects to refine the good governance outcome indicators such as those illustrated in Table 3.

In addition, we believe that the proposed framework constitutes a relatively sound research approach to act as a good starting point for empirical studies on the assessment of the effects or benefits of e-Governance. Given the limited knowledge in this area, we invite researchers to use and test the proposed outcome evaluation framework and use case studies to build the required knowledge in this area.

Endnotes

1 Gartner Symposium/ITxpo, held between April 29 and May 2 at the San Diego Convention Center, San Diego, USA. Available at: http://www.gartner.com/5_about/press_releases/2002_04/pr20020430b.jsp

2 LOG-IN Africa is a pan-African network of researchers and research institutions from Egypt, Ethiopia, Kenya, Mauritius, Morocco, Mozambique, Senegal, South

Africa and Uganda. The Network has been carrying out projects to assess the state and outcomes of e-local governance (or the application of ICTs to transform the business of local governments and create governance outcomes) and initiatives. The overall research question is: *"What progress has been made and what are the outcomes in the provision of e-local governance in Africa? What are the challenges and threats? And, what are the good practice strategies and solutions that are emerging?"* More details can be found at:http://www.loginafrica.net

3 The World Bank Group (2006), 'What is Governance? Arriving at a common understanding of 'governance', Available at: http://www1.worldbank.org/mena/governance/issues-keyQuestion.htm.

4 The Urban Governance Initiative (TUGI) 'Good Governance Issues Report Card: Solid Waste,' Available at: http://www.tugi.org/reportcards/solidwaste.PDF.

5 The other major outcome categories were technology, organization, citizen and regulation.

6 The research work done by Kettani *et al.* (2005) at Fez City in Morocco was a key basis for the LOG-IN African project. The work by the same researchers in the City of Larache in Morocco, based on the work in Fez, is one of the projects being carried out under LOG-IN Africa.

7 LAIFOMS is a computer-based system that integrates financial and operational activities involving business transactions in a Local Authority. The system has three main components, namely: revenue, budgeting, and financial management and expenditure components. Each of the components is integrated with each other to have a comprehensive system to monitor all operational activities.

8 In this illustration, the "e-society" e-Government dimension is only applicable to the "participation" good governance construct.

References

ADB (2003), 'To serve and to preserve: Improving public administration in a competitive world,' Available at: http://www.adb.org/documents/manuals/serve_and_preserve/Chapter19.pdf.

Heeks R. (2003), 'Most e-government-for-development projects fail: How can risks be reduced?,' *IDPM i-Government Working Paper no.14.*, University of Manchester, UK, Available at: http://idpm.man.ac.uk/publications/wp/igov/index.shtml.

Kettani D., Moulin B., and Elmahdi A. (2005), 'Towards a formal framework of impact assessment of E-Government systems on governance,' *Conference Proceedings, the Fourth WSEAS International Conference on E-ACTIVITIES* (E-ACTIVITIES '05), Miami, Florida, USA, pp. 12-20.

Kusek J. Z. and Rist J.C. (2004), Ten steps to a Results-Based Monitoring and Evaluation System: A Handbook for Development Practitioners, Washington D.C.: The World Bank.

Mayne J. (2007), 'Challenges and Lessons in Implementing Results-Based Management,' in *Evaluation*) 87-109.

Ndou V. (2004), 'E-Government for Developing Countries: Opportunities and Challenges,' in *The Electronic Journal on Information Systems in Developing Countries*, 18 (1): 1-24, Available at: http://ejisdc.orgOECD (2003), 'The e-Government imperative: Main Findings,' in *OECD Observer*, March, Available at: http://www1.oecd.org/publications/e-book/4203071E.PDF.

Pacific-Council (2002), *Roadmap for e-Government in the developing world: 10 questions E-Government leaders should ask themselves*, Los Angeles, CA: Pacific Council on International Policy. Available at: http://www.pacificcouncil.org/pdfs/e-gov.paper.f.pdf.

Republic of Kenya (ROK) (1999), Report of the rationalization and staff rightsizing for effective operation of the Ministry of Local Government, Nairobi: Government Printer.

Rockcliffe-King and Mitullah W. V. (2003), Support to Local Government Programme Components: Output to Purpose Review, Nairobi: DFID.

Saxena K. B. C. (2005), 'Towards excellence in e-governance,' in *International Journal of Public Sector Management,* 18(6): 498-513.

Seifert J. W. and Bonham G. M. (2003), 'The transformative potential of e-Government in transitional democracies,' Available at: http://www.mawell.syr.edu/maxpages/faculty/gmbonham/Transformative_Potential_of_E-Government.pdf.

The Global Development Research Center (GDRC) (2004a), 'Urban Governance,' Available at: http://www.gdrc.org/u-gov/index.html.

The Global Development Research Center (2004b), 'Understanding the Concept of Governance,' Available at: http://www.gdrc.org/u-gov/governance-understand.html.

The Global Development Research Center (2004c), 'Governance: A Working Definition,' Available at: http://www.gdrc.org/u-gov/work-def.html.

UNDP (2002), *Handbook on Monitoring and Evaluation of Results. UNDP Evaluation Office*, Available at: http://stone.undp.org/undpweb/eo/evalnet/docstore3/yellowbook/documents/full_draft.pdf (last accessed 26 June, 2007).

Waema T. M. (2005), 'A Brief History of the Development of ICT Policy in Kenya,' in Etta F. and Elder L. (eds.),*At the Crossroads: ICT Policy Making in East Africa*, Nairobi, Kenya: East African Educational Publishers Ltd.

Waema T. M. and Mitullah W. (2007), 'E-Governance and Governance: A Case Study of the Assessment of the Effects of Integrated Financial Management System on Good Governance in Two Municipal Councils in Kenya,' *ACM Proceedings of the 1st International Conference on Theory and Practice of Electronic Governance*, Macao, China, 10-13 December, pp. 263-268.

Weiss T. G. (2000), 'Governance, Good Governance and Global Governance: Conceptual and Actual Challenges,' in *Third World Quarterly*, 21(5): 795-814.

Chapter 7

Policy Implications of the Relationship between ICT Access and Usage and Well-being:

A Case Study of Kenya

Timothy Mwololo Waema & Obadia Okinda

1. Introduction

Kenya has witnessed increased penetration of various ICTs, including mobile telephone, internet, radio and television. This has had the effect of these technologies reaching a greater proportion of the population, including the poor. The Government of Kenya continues to view ICT as a powerful tool in accelerating productivity of all sectors and empowering people to meet the challenges of the 21st century. To optimize the sector's contribution to the development of the entire economy, the government is currently offering investment opportunities to the private sector. Moreover, the ICT sector has been identified as one of the areas to contribute to economic growth, with Business Process Off-shoring (BPO) aimed at creating jobs and contributing to GDP growth (Government of Kenya, 2008b). ICTs have also been used to enhance provision of public services important in poverty reduction through such initiatives as e-government.

At a time when expansion of ICTs is rapid, coupled with huge investments by government and the private sector, it is crucial to examine the relationship between ICTs and poverty reduction. The idea that ICTs can act as facilitating factors linking social, economic and natural well-being by improving communications and networking, whether social or economic in intention, and by reducing exclusion through information processing and dissemination, the question of poverty reduction-ICT nexus especially at a micro level needs to be adequately answered. This is important in Kenya where poverty reduction remains a major concern and ICTs now feature in key development policy documents, unlike in the 1970s, the 1980s, and the 1990s.

Some researchers argue that the role that ICTs play in poverty reduction is overestimated with a lot of emphasis on the supply of ICTs (for example, Sreekumar and Rivera-Sánchez, 2008; Tiwari, 2008). They further point out that there is little empirical evidence of the role of ICTs in poverty reduction especially at a micro level (household and individuals). In addition, there is minimal evidence that the relative benefits of investment in ICT infrastructure would compare with benefits from investment in education, health, roads and dams, and industrial parks (Leye, 2009). Hence, a positive relationship between ICTs and poverty reduction either is weak or remains unclear and evidence from ICT projects is problematic and isolated (Wade, 2000; Adeya, 2002).

It is therefore clear from the literature that the nature of the relationship between ICTs and poverty remains unclear and research to date has resulted in divergent conclusions. This divergence in understanding the relationship between poverty and ICTs appears to be caused in part by differences in terms of underlying causes of poverty and how poverty reduction can be attained. In some empirical ICT studies, the relationship between ICTs and poverty is arrived at by analysing the potential impact of ICTs on macro-economic variables or evaluating ICT-centred interventions (Akpan-Obong, 2010). This analysis tends to ignore the effect of ICT access and usage among households and individuals as they go about their day-to-day socio-economic and political activities. Moreover, many evaluations of the use of ICTs have studied impacts of ICT projects on people's livelihoods (Soriano, 2007; Tiwari, 2008). These information deficiencies have raised concerns among policy makers who are being repeatedly urged to invest a substantial part of the national budget in ICT infrastructure based on an incomplete evidence base. In this situation, it is tempting to question whether investment in ICTs represents the best option for poor communities.

The aim of the study was to investigate whether and how access and use of ICTs by households improves their well-being or poverty reduction in order to guide policy interventions.

2. Conceptual Framework

To analyse the relationship between ICTs and poverty, this study utilises the livelihoods approach. Duncombe (2007) has extended and used the livelihoods approach to analyse ICT applications for poverty reduction through the micro enterprise. This approach is appropriate for this study because it incorporates key variables relating to people's livelihoods and environmental context whose relationship is sought by this study. A livelihood is conceived as the activities that people utilize to earn a living and the capabilities and assets (material and social) that contribute to or influence their ability to earn a living. The livelihood approach takes a systems view which recognises that livelihoods operate within wider social, political and economic context (DFID, 2000). The livelihoods framework (see Figure 1) posits that although the poor are in a vulnerable context, they have access to certain assets or capitals that they utilize to improve their well-being. However, the poor pursue their livelihood objectives within the prevailing institutional, organizational and social environment which influences the extent to which their social and economic objectives can be attained. This framework therefore incorporates key variables of interest to this study.

Figure 1: The Livelihoods Framework

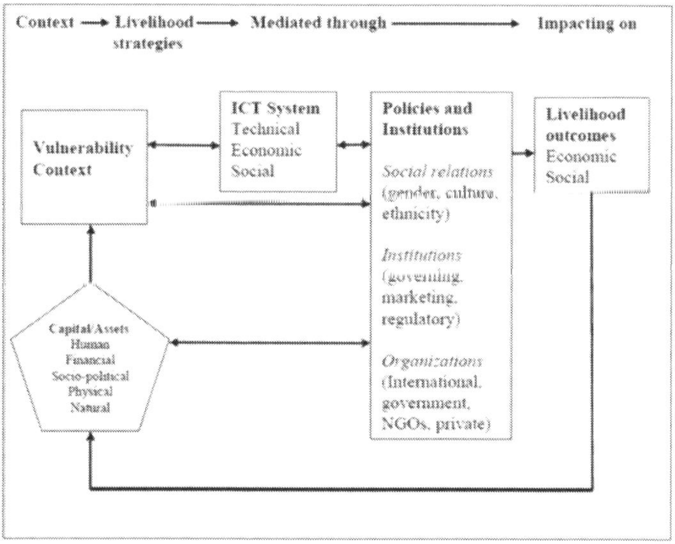

Source: Adapted from Duncombe (2007)

The focus is thus on people's livelihoods, the institutional context within which the poor pursue their livelihood objectives, the outcomes of their livelihood activities and how these outcomes influence their assets and vulnerability context. The ICT system which incorporates various ICTs is conceptualised to influence people's livelihood strategies within the wider policy and institutional context. Thus, livelihood outcomes (social and economic) of households and individuals are affected by the ICT system and the wider policy and institutional context. The outcomes in turn influence peoples' assets, which would in turn change their vulnerability context (feedback loop).

This study looks at poverty from a multidimensional perspective as consisting of economic (income, consumption, assets) and social (access to services and infrastructure, social inclusion/exclusion, empowerment, risk and vulnerability) elements. The study combines the social and human assets as the social aspect, and physical and financial capitals are combined as the economic aspect. It is thought that these measures are sensitive to change induced by use of ICTs. In addition, these social and economic outcomes could influence people's capitals/assets which would in turn change their vulnerability context. Poverty reduction is viewed as improvement in people's livelihood outcomes (social and economic). The economic dimension looks at those elements that relate to the income or consumption measured by the conventional money-metric indicators like per adult equivalent household consumption, income, and the financial and physical assets like land. On the other hand, the social dimension includes the notion of access to essential services and infrastructure, social exclusion/inclusion, empowerment, participation and the incidence of risk and vulnerability and the capacity to cope with, or adapt to such vulnerability.

ICTs are broadly defined to incorporate the complete range of communication, media and information flows in a community including radio and television; telephony including mobile, fixed line; and Internet access. Thus, ICTs and their usage is influenced by not just the technical factors (end user technologies, networks and access to infrastructure and applications used), but also by the economic context (economic institutions, cost structures and regulatory framework) and social context (social actors, political processes, social interactions and networks and the content). The economic and social sub-systems influence the household access and usage of the ICT technologies in the technical sub-system to affect the various dimensions of poverty. Analysis in this study sees ICTs as tools that enhance poor people's livelihood strategies within a particular institutional and policy context. The study analyses how people's access and use of ICTs changes their social and economic livelihood outcomes.

3. Research Methodology

The core of the methodology for this study was a panel/longitudinal research consisting of a quantitative study at the start of the research project repeated about 24 months later using the same households. The 'before and after' periods was to be used as a window to assess the impact of ICTs and to find out whether there is a relationship between the level of use of these facilities and changes in poverty indicators. The aim was to compare and contrast changes in the use of ICTs and in poverty levels and to ascertain the impact through in-depth interviews with the households where the change was significant or insignificant or totally absent.

The first wave was a survey carried out using a questionnaire towards the end of 2007 and completed in early 2008. Sampling was from within Research ICT Africa (RIA) sample of July 2007[1], which utilized the NASSEP IV sampling frame of the Kenya National Bureau of Statistics (KNBS). The purposively predetermined sample size of 400 was evenly spread across the 20 Enumeration Areas (EAs) purposively determined. Six EAs were sampled from the 16 that comprised the poor under major urban areas, 2 EAs were from each of the selected three municipalities, one each from lower middle poor (category 4) and lower poor (category 5) of the KNBS classification.[2] In addition, 14 EAs were selected from the fourth quartile of 9 districts classified as the poorest within the RIA sample.

The second quantitative study carried out in 2010 involved interviews with households covered during the first wave in 2007. Fieldwork therefore targeted the 400 households covered in 2007 as well as 400 individuals interviewed from these households. The research team however only located and interviewed 304 households and 288 individuals. The major reason for the failure to locate some households was that households had relocated and the research team could not obtain any information through which to contact them.

Both waves of quantitative data collection were complimented by qualitative data collection through focus groups discussions, key informant interviews and case histories using interview schedules. Respondents were selected purposively from a sample of EAs out of the 20 that were covered during the quantitative surveys.

The demographic data were used to calculate the household size and the corresponding adult equivalence in each household while consumption data were used to calculate the total consumption aggregates. These were food and non-food consumption aggregates. The food component (which contained 44 food items) was collected within a period of one month. The non-food component had two recall periods of a month and one year.

Monthly per adult equivalent for both total food consumption and total non-food consumption was used to calculate household welfare measures that were used to classify households into poor and non-poor groups. Two cases in the 2007 dataset did not have adequate consumption data while 11 cases in the 2010 data had no consumption data and were dropped in the analysis of household well-being. The Cost of Basic Needs (CBN) approach was used to determine poverty lines. This approach requires that the calorific value of the food items for a given basket of commonly consumed food items by all members of the community/society be established and used. A basket developed by KNBS using the Kenya Integrated Household Budget Survey (KIHBS) dataset and costing the basket using the national Consumer Price Index (CPI) prices corresponding to the months that the data were collected. Appendices 1 to 4 show food poverty line derivation for the respective months of data collection.

Absolute poverty line was obtained by adding food poverty line to the non-food component. Non-food component was obtained by computing the mean value of total non-food expenditure consumed by households whose food expenditure falls within a one percentage point interval around the food poverty line. This process was repeated five times and at each stage the interval was increased by additional two percentage points. The average of the mean total non-food expenditure from each stage provides a weighted non-parametric estimate of the value of the non-food component which was added to the food poverty line to compute the overall poverty line. The overall poverty lines in monthly adult equivalent terms were computed as KShs 2173.1, KShs 2398.9, KShs 2503.1 and KShs 3061.3 for December 2007, March 2008, April 2008 and February/March 2010, respectively.

The main challenge was in the second wave of data collection. It was difficult for the field guides to trace and locate households and individual ICT respondents.

4. Existing Policy Situation

Poverty reduction policy

Perhaps in recognition of weaknesses that befell previous poverty policies and the continued increase in population, the government came up with a comprehensive development blueprint dubbed the 'Kenya Vision 2030' in 2008. This policy stipulates broad economic, social and political reforms and programmes to be implemented to, among others, attain high levels of economic growth and reduce poverty and impoverishment. The social sector of the vision aims to reduce the number of absolutely poor people, enhance people's access to public and basic services and promote equity. The first five-year Medium-Term Plans (MTP) 2008-2012 seeks to attain a high economic growth rate and increased employment opportunities to address high unemployment, large number of poor people, and (income, gender and geographical) inequality (Government of Kenya, 2008b).

Although it is early to evaluate the progress of the Vision 2030 and the first MTP (2008-2012) in poverty reduction, it is argued that the extent to which the vision will lead to poverty reduction depends on how it addresses among others, structural and political processes that cause poverty and challenges that have stalled the country's development. A report by KIPPRA (2009) notes that Kenya's growth target set by Vision 2030 of a double digit annually is ambitious and cannot be realized and sustained without a keen commitment to reforms. Further, the challenges to Kenya's development identified in 2005 by NEPAD Africa Peer Review Mechanism (APRM) that included managing national diversity, transformative leadership, implementing national development plans and strategies, poverty and inequality of wealth distribution, land, corruption, gender equity, and youth unemployment have not been adequately addressed.

There is a lot of optimism that reforms outlined in the Vision 2030 and the newly promulgated constitution that focuses on economic and political institutions will be the much needed impetus to the country's development and poverty reduction. Regarding the role of political reforms, Leftwich (2008:4) notes that "Reduction of poverty in recent and contemporary history has generally been attributable to two processes: Economic growth which creates opportunities and jobs, and political processes which develop and sustain institutional arrangements that provide both safety nets and redistributive provision (welfare regimes, broadly)." It has also been recognized that on its own, rapid market-driven, or even state-driven, transformative economic growth can leave many people or regions out of the process and generate profound inequalities. To this end, it is yet to be seen how the implementation of the new constitution and the rest of the Agenda 4 activities[3] will reform the nature of the country's politics, the basis of which any sustainable economic development and poverty reduction can be attained.

ICT policy

The current national ICT policy is the first for the country and was published through a special issue of the Kenya Government Gazette (Government of Kenya, 2006). The vision in this policy is of "a prosperous ICT-driven Kenyan society". Its mission is "to improve the livelihoods of Kenyans by ensuring the availability of accessible, efficient, reliable and affordable ICT services". This policy needs review given the changes that have taken place in the last five years and the changes that are anticipated. For example, there is a converged legal and regulatory environment, the undersea fibre bandwidth has landed and is operational, there is Vision 2030 that the policy needs to align to, there is a new constitutional dispensation and there have been and will be many other national, regional and global changes.

The key component the national ICT policy that is pro-poor is the section that addresses universal access. Here, the government commits to ensuring that ICT services are available throughout the country and to support the universal service availability and the widest access to such services. In order to finance universal access, the Government commits to promote investments in rural and other un-served areas through provision of incentives. The government also commits to set up Universal Service Fund for the ICT sector to complement private sector initiatives towards meeting universal access targets. This Fund would subsidize the cost of ICT infrastructure rollout and expansion to un-served and under-served areas by operators in the sector. The Fund would also target training in ICT services.

One key initiative is the landing of the undersea cables – Seacom and The East African Marine System (TEAMS) – in 2009 have already had a dramatic effect. The Eastern Africa Submarine Cable System (EASSy) landed in 2010. The landing of the cables brought a lot of consumer expectations but only quality had improved and by the end of this study, the reduction in the cost of retail Internet service remained elusive. The Kenyan Government has also invested in building a national fibre backbone, referred to as the National Optic Fibre Backbone Infrastructure (NOFBI). The final initiative is by the Ministry of Information and Communication to develop rural ICT centres ("pasha") to ensure appreciation and use of ICT throughout the country. This initiative will contribute towards universal access/services.

The vision of the ICT sector MTP of Vision 2030 is that "Kenya becomes an information and knowledge based society". Its mission "is to provide information and communication based solutions and ensure the availability of efficient, reliable and affordable information communication services countrywide". The overall sectoral goal of the ICT sector MTP is "to facilitate provision of equitable and affordable quality information and communication services countrywide" (Government of Kenya, 2008a). In addition, there is a draft National ICT Master Plan. Two of the key pro-poor strategies in this plan is to implement initiatives to ensure universal access to broadband services and to make e-government services universally available as part of service delivery to the citizen.

The most influential document regarding ICT legislation and regulation in Kenya has been the recently amended Kenya Communications

Act 1998 (Kenya Gazette Supplement, 1998). One of the key sections in the new law, the Kenya Communications (Amendment) Act 2009 (Kenya Gazette Supplement, 2009), is Part VII on electronic transactions (e-transactions). In this part, the new Act, among other things, gives legal recognition of electronic records; recognizes electronic messages as valid for the formation of contracts; and supports the use of electronic records and electronic signatures in government and its agencies. The new Act also deals with various aspects of ICT and cyber-crimes.

Another key feature of the Kenya Communications (Amendment) Act, 2009, is the establishment of a Universal Service Fund (USF) to support widespread access to, support capacity building and promote innovation in ICT services. The creation of the USF encourages equal access to ICT services which enhances information access. Finally, there is the pending Freedom of Information Bill, which if enacted, will, among other things, enable the public to access information in the possession of the government and public authorities and certain private bodies that have a public character.

5. Findings

Access and usage of ICTs

Although radio was by far the most widely accessed and used ICT, it had been overtaken by mobile telephone while internet penetration remained low though its use was increasing. Perhaps one of the explanations for this is convergence of technology - some mobile phone handsets have radio, internet and other applications. The low household access to internet and fixed telephony compared to mobile phones is borne by national statistics. For example, the 2009 national statistics show mobile phone penetration of 44% by 2009, a total of 3,648,406internet users by June 2009 representing a penetration of less than 10%, and less than 1% penetration of fixed telephony (Waema et al., 2010).

The key reasons for high access to radio and mobile phones were that the two ICTs were more affordable, more suitable for a small house, did not require electricity in the house and were less susceptible to theft. This has implications for promotion of other ICTs like computers due to safety reasons in areas considered insecure. There was also evidence that ICT use was higher than access (ownership) especially for telephone (fixed, mobile), radio and television due to sharing among friends and relatives as well as use of public ICTs.

In summary, the study found that the key factors affecting ICT access and usage were:

- Geographic location: access and usage of ICTs remains largely an urban phenomenon. This means that there is a 'digital divide' between urban and rural areas with respect to access to ICTs;
- Household headship: Heads of households are more likely to have mobile phones compared to other members of their household;
- Marital status: Married/cohabitating persons are more likely to own mobile phones compared to those who are divorced, separated, widowed or single;
- Occupation: Although mobile phone ownership is higher among the employed, even the unemployed own mobile phones. Notable is also that among those who had an email address, the proportion of the employed is close to that of the unemployed;
- Education: People with higher levels of education are more likely to have email addresses and mobile phones than those with lower levels of education;
- Gender: There was no statistically significant difference between gender and mobile phone ownership. Thus, mobile phone ownership does not vary by gender.;
- Age: Email addresses were highest in the 15-29 age group, which means that the youth are more likely to have internet access.

In comparison to other countries, Kenya had the highest access to most digital ICTs as shown in Figure 2. Although Rwanda seems to have higher e-mail access, national figures show that Kenya has the highest Internet penetration in the region (see Waema et al., 2010).

Figure 2: Comparison of percentage HH access to ICTs in East African countries

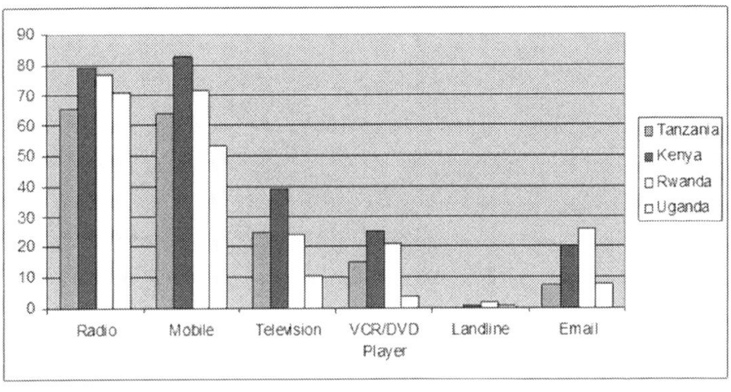

Source: Study, May 2010

ICTs and changes in different dimensions of poverty

The key findings on changes on the various dimensions of poverty (financial poverty, vulnerability, human capital, asset poverty, physical poverty, social exclusion/inclusion, and digital poverty) over the period between 2007/08 and 2010 are outlined below.

Financial poverty

The quantitative analysis revealed that the levels of absolute poverty increased over the period 2007 to 2010. Table 1 shows poverty dynamics (movements into and out of poverty) over the period. Financial poverty here is in terms of per adult household expenditure.

Table 1: Poverty transitions based on expenditure

Poverty status 2007	**Poverty status 2010**		**Total**
	Poor	**Non Poor**	
Poor	63	24	87
Non poor	68	144	212
Total	131	168	299

Source: Field data, 2007/08; 2010

This could largely be explained by the country's declining economic performance over the period. The country recorded increased economic growth rates from 2003 that peaked in 2007, with real GDP growth peaking at 7.1%. However, due to the disruption following the disputed 2007 general election and poor weather that led to the contraction of the agricultural sector, the economy performed poorly in 2008 (1.6%) and 2009 (2.6%).

Despite this negative relationship, qualitative analysis found evidence of increases in income due to enhanced use of ICTs resulting from gains in efficiency, effectiveness, and savings from operational costs in income generating activities. In addition, ICTs were found to have a positive effect on travel cost. Increased use of ICTs, especially mobile phones, meant that people did not have to travel at all or only travelled for short distances. This was achieved through calls made to obtain or send information instead of travelling, thereby saving on travel time and costs.

Finally, quantitative analysis showed that mobile money transfer was an additional mechanism the poor used to reduce costs of travel and sending money. The findings showed that a quarter of the households surveyed received or sent money (26% and 23%, respectively). The findings also show that poor households mostly received money from someone working in another village or city while non-poor households received money mostly from someone who lived abroad/out of the country. In terms of the channels used, ICT modes particularly mobile money transfer services (Mpesa and Zap-Airtel money) were the third most common channel used after traditional banking modes and hand delivery modes. Mobile phone calls were the most common means of informing households that money had been transferred, 51% of household that received and 59% of households that sent. Transfer of money by poverty category of households is shown in Figure 3.

Figure 3: Money transfer by poverty category (%)

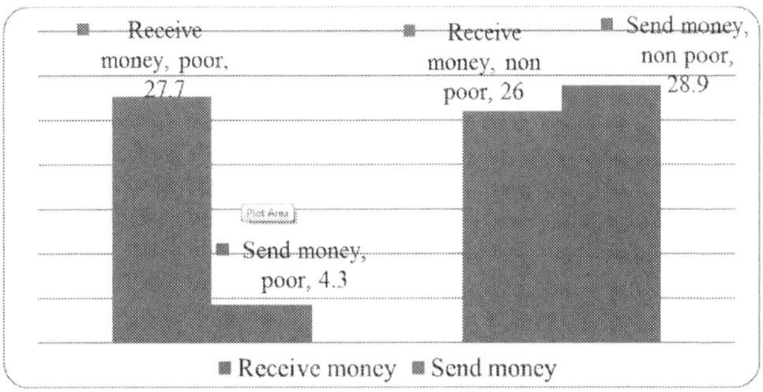

Source: Field data, 2007/08

Figure 3 shows that a higher percentage of non-poor households send money compared to that of poor households. In terms of association, the relationship is significant for sending money ($p<0.005$) but not for receiving money ($p>0.005$). This implies that while non-poor households were more likely to send money compared to poor households, both categories of households were equally likely to receive money.

Asset poverty

There was a statistically significant increase in the sum of assets (refrigerator, electric cooker or gas stove, television set, lounge suite, motor vehicle and livestock).The mean sum of household assets increased from 2.28 in 2007 to 2.92 in 2010.

Physical poverty

Access to services and infrastructure, that is the sum of waste removal (protected or unprotected), access to running water, sewerage/toilet, electricity, and the structural condition of a dwelling unit (condition of walls and roof), statistically improved from 2.77 in 2007 to 3.46 in 2010.

Vulnerability or shocks

Households used ICTs in coping with vulnerabilities. Vulnerability was measured in terms of shocks that households experienced between 2005 and 2007/08 when the study was conducted. About half of the households covered reported to have experienced a shock. Some households experienced more than one shock and of the 14 shocks mentioned, the three most common were serious injury or illness (21%), theft/fire/destruction of household property (12%) and death/disease of livestock (8%). Figure 4 illustrates use of mobile phone by households to assist them during the two most common shocks - serious injury or illness, and theft/fire/destruction of household property.

Figure 4: Use of ICTs during negative shocks by poverty category

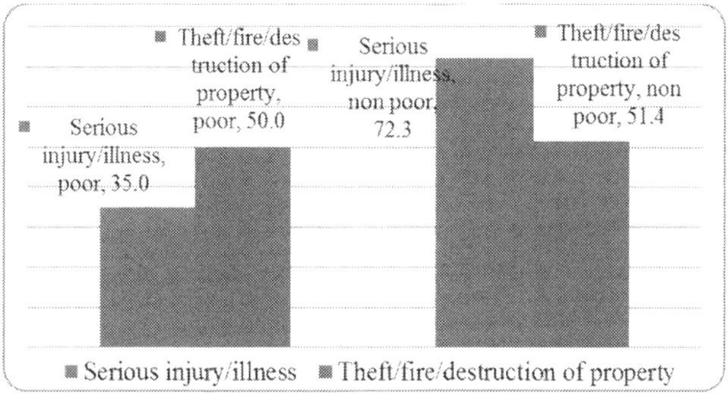

Source: Field data, 2007/08

Generally, all categories of households used a mobile phone to assist them during these two shocks. Further, use of mobile phone during serious injury or illness seems to vary between poor and non-poor households (higher). This is supported by chi square test that shows association between the two variables (p<0.005). However, use of mobile phone during theft, fire or destruction of property seems to be similar (p>0.005). This implies that households, including those considered poor used mobile phone to assist them during theft/fire/destruction of property.

This finding is supported by findings from qualitative survey which showed that some households, including those classified as poor, used internet and mobile phones to assist them when they experienced serious injury or illness of household members and theft, fire, or destruction of household property and used radio and mobile phone to get jobs. The qualitative study also established that use of ICTs had enabled people to seek help during emergencies like security problems; to know what health services were available and where; to report health problems faster; and to obtain informative health information important to their livelihoods.

Social inclusion/exclusion (participation in social networks)

Although the mean sum of people's participation in groups and social networks decreased from 1.37 in 2007 to 1.34 in 2010, this decrease was not statistically significant. This was corroborated by the qualitative survey which showed that there were people who belonged to groups that did not use ICTs in their activities; and that ICTs did not influence them to join these groups. The qualitative study however found that enhanced use of ICTs seemed to have improved the activities of economic and social groups that were already in place. In particular, some groups used mobile phones to transfer money among members and successfully and conveniently set up group meetings.

Human capital

Human capital (change in the highest level of education attained by household members) increased from 1.82 in 2007 to 1.86 in 2010, an increase that was statistically significant. This was corroborated by the qualitative study which established that ICT had positively influenced human capital (less poverty) as measured by the change in the highest level of education attained by household members. From the qualitative study, use of ICTs especially radio, internet (mobile phone and computers) and television was associated with the acquisition of knowledge and skills among household members. Thus, use of ICTs led to improved farming practices, as well as access to education information for those with informal school education.

Despite the above effects of access and usage of ICTs on different dimensions of poverty, the qualitative survey found that greater usage and impact of ICTs on households was limited by the high cost of purchase, operational and maintenance of ICTs; lack of or unreliable power supply; poor mobile network coverage and quality of services; and inadequate skills to use some of the ICTs, especially computers and the internet.

5. Conclusions

The study has shown mixed results on the effect of access and usage of ICT with livelihoods (or poverty) at the micro (household and individual) level. Whether ICT has a positive or negative influence on household livelihoods depends on other factors including macro-level contexts. It is also possible that micro-level contexts can also influence the impact of ICT usage on household livelihoods. For example, there was evidence that changes in ICT access and usage are not uniform. Access and usage of some ICTs increased while it decreased for other ICTs, even for persons in the same household. This is due to the fact that households differ in what ICT they consider important in their activities and choose what they should access and use. For example, mobile money transfer had increased the use of mobile phones while access to educational, agricultural, health, political and other valuable information through the radio and television had increased the use of these technologies.

Therefore it can be argued that households are impacted differently by increased usage of ICTs depending on factors internal and external to the household. These factors are what researchers (such as Zheng and Walsham, 2007) who have carried out ICT studies using Sen's Capability Approach have referred to as Conversion Factors. This contextual view of the impact of ICTs on poverty is what May alludes to when he avers that "... it seems probable that the impact of ICT will be determined by the context in which these technologies are deployed, the preparedness of the users and the opportunities that exist for their application" (May, 2010). It is also in line with the social informatics perspective of information systems (Kling, 2000) where the social context is critical in determining the impact of information systems in organizations, and by extension, in societies.

Based on the findings, mobile phones and radio have the most potential for poverty reduction in Kenya due to their high levels of access and usage. Internet penetration remains low especially in rural areas due to lack of appropriate infrastructure and requisite knowledge among the people – although this is rapidly changing with increased use of mobile internet. Television remains largely pervasive in urban areas. Thus, the discussion of ICTs should not just be about internet and computers because the 'traditional ones' like radio still play a critical role.

Although ICTs were found to have benefits that had positive effects on household livelihoods, these benefits are however limited by challenges associated with greater usage of ICTs by households. These include the high cost of purchase, high operational and maintenance cost, lack of or unreliable power supply, poor mobile network coverage and inadequate skills to use some of the ICTs, especially computers and the internet.

6. Policy interventions

Cost still remains the most important factor limiting ownership, access and usage of key ICTs. We therefore recommend policy and regulatory interventions to reduce the cost of ownership, access and usage of ICTs by the consumer. Key here is the reduction of the retail cost of broadband. Such interventions should aim at enhancing universal access and utilisation, ensuring equity along urban/rural and poor/non-poor households as well as gender dimensions. This can be achieved through fast-tracking the implementation of the universal access through utilizing the universal service fund created by the Kenya Communications (Amendment) Act, 2009. Universal access implementation should address the 'digital divide' between urban and rural areas, and will require an innovative combination of conventional (e.g. radio) and new (e.g. mobile) technologies as well as some new forms of ICT business models. Universal access implementation should also be accompanied by an accelerated implementation of the rural electrification program. If universal access strategy is not implemented, ICT impacts on development in general and on poverty will remain limited.

Secondly, it is important to articulate the role of ICTs in poverty reduction in key development policy documents to ensure that development efforts and investments are directed to areas that matter to the well-being of the poor. This is essential due to the rapid expansion of the ICT sector characterised by increasing usage of ICTs for diverse purposes by households.

The third recommendation is that the Ministry of Information and Communications should lead the review of the proposed Freedom of Information Bill in line with the new constitutional dispensation which makes access to information held by the state or a person a fundamental right for every citizen. Further and in line with article 35 of the new constitution on access to information, the Directorate of e-Government should roll-out e-government projects that ensure that information is provided to the poor and that information is available to inform the policies, institutions and processes that affect the poor. If this is not done, the Government might end up with a plethora of law suits by citizens on access to information.

The level of literacy and capacity to use ICT systems is one of the major factors that influence adoption and usage. A further recommendation is therefore that the Government should facilitate, through participation of the private and civil sectors as necessary, the enhancement of people's ICT skills and awareness of the potential of various existing and future ICTs on their well-being. It will also be necessary to introduce a relevant ICT curriculum at basic levels of education (primary and secondary schools). If this is not done, adoption and usage of ICTs, especially in rural areas, will remain limited and so will the impact.

There is still rampant theft and robbery that is associated with ICTs, such as the theft that takes place though mobile phones and cyber-crimes and malpractices through the use of the internet. This increases the vulnerability of poor people with ICTs. The final recommendation is therefore that the appropriate state security organs and the judiciary should enforce compliance to the Kenya Communications (Amendment) Act, 2009, which spells out cyber-crimes and the associated punishment. If this is not done, the gains from impacts of ICTs will quickly be eroded.

Notes

1 Research ICT Africa (RIA) survey, conducted in Kenya in 2007 investigated ICT access and usage.

2 KNBS classifies clusters into five categories or classes: 1. Upper Class; 2. Upper Middle Class; 3 Middle Class; 4. Lower Middle Class; and 5. Lower Class.

3 These are the long-term issues and solutions that were agreed upon after Kenya's disputed 2007 elections by all the parties in the Kenya National Dialogue and Reconciliation that was chaired by H.E. Kofi Annan, Chair of the Panel of Eminent African Personalities.

References

Adeya, C. N. (2002), *ICTs and Poverty: A Literature Review*, Ottawa: IDRC.

Akpan-Obong, P. (2010), 'Unintended Outcomes in Information and Communication Technology Adoption: A Micro-level Analysis of Usage in Context,' in *Journal of Asian and African Studies*, 45 (2): 181-195.

DFID (2000), *Sustainable Rural Livelihoods Guidance Sheet*, London: Department for International Development.

Duncombe, R. (2007), 'Using the Livelihoods Framework to Analyse ICT Applications for Poverty Reduction through Micro enterprise,' in *Information Technologies and International Development*, 3(3): 81-100.

Government of Kenya (2008a), *Information and Communication Technology (ICT) Sector Medium Term Plan for Kenya Vision 2030*; 2008/09-2011/12, Ministry of Information and Communications, Nairobi: Government Printer.

Government of Kenya (2008b), *Kenya Vision 2030: First Medium Term Plan (2008-2012)*, Ministry of State for Planning, National Development and Vision 2030, Nairobi: Government Printer.

Government of Kenya (2006), *Information and Communications Technology Sector Policy Guidelines*, Gazette Notice No. 2431, Nairobi: Government Printer.

Kenya Gazette Supplement (2009), *The Kenya Communications (Amendment) Act No. 1 of 2009*, Kenya Gazette Supplement, Nairobi: Government Printer.

Kenya Gazette Supplement (1998), *The Kenya Communications Act, 1998 and The Postal Corporation Act, 1998*, Supplement No. 64 (Act No. 3) Special Issue, Kenya Gazette Supplement, Nairobi: Government Printer.

KIPPRA (2009), *Kenya Economic Report 2009: Building a Globally Competitive Economy*, KIPPRA: Nairobi.

Kling, B. (2000), 'Learning about Information Technologies and Social Change: The Contribution of Social Informatics,' in *The Information Society*, 16(3): 217-232.

Leye, V. (2009),'Information and Communication Technologies for Development: A Critical Perspective', in *Global Governance*, 15 (1): 29-35.

Leftwich, A. (2008), *Developmental States, Effective States and Poverty Reduction: The Primacy of Politics*, UNRISD Project on Poverty Reduction and Policy Regimes,UNRISD: Geneva.

May, J. (2010), 'Digital and other poverties: Exploring the connection in four East African countries', *Paper presented at the conference - Ten years of 'war against poverty': What we have learned since 2000. What we should do 2010-2020*, 8-10 September,Manchester, UK.

Soriano, C. R. (2007), 'Exploring the ICT and Rural Poverty Reduction Link: Community Telecenters and Rural Livelihoods in Wu'an, China,' *EJISDC*, 32 (1): 1-15.

Sreekumar, T.T. and Rivera-Sánchez, M. (2008), 'ICTs and Development: Revisiting the Asian Experience,' in *Science, Technology & Society*, 13(2): 159-174.

Tiwari, M. (2008), 'ICTs and poverty reduction: User perspective study of rural Madhya Pradesh, India,' in *The European Journal of Development Research*, 20(3): 448-461.

Wade, R.H. (2000), 'Bridging the Digital Divide: New Route to Development or New Form of Dependency?,' in *Global Governance*, 8(4): 443-66.

Waema, T., Adeya, C. and Ndung'u, M.N. (2010), *Kenya Draft Sector Performance Review 2009*, Available at: http://www.researchictafrica.net.

Zheng, Y. and Walsham, G. (2007), 'Inequality of What? Social Exclusion in the e-Society as Capability Deprivation,' *Working Paper no.167*, Information Systems Dept, London School of Economics, London.

Appendices

Appendix 1: Poverty Derivation for Dec 2007

Food item	Share in basket	Kcal (100g)	CPI price (Kshs./100g)	Kcal per Kshs. 100	KShs Per 2,250 Kcal
Beef with Bones	0.09	156	18.4	76.4	4.9
Sugar	0.08	375	7.0	418.8	4.3
Milk Fresh Unpacked	0.07	72	6.0	81.8	3.7
Bread	0.07	257	7.8	219.8	3.6
Maize Flour Sifted	0.06	264	2.4	652.0	3.3
Milk (Fresh Packed)	0.05	72	6.0	54.9	2.5
Tomatoes	0.04	27	3.9	30.2	2.4
Rice (Grade 2)	0.04	346	5.1	270.0	2.2
Maize Grain Loose	0.04	353	1.9	670.5	2.0
Kale (Sukuma Wiki)	0.04	52	2.2	83.2	1.9
Beans	0.03	314	4.9	211.0	1.8
Cooking Fat	0.03	900	15.9	81.6	1.8
Potatoes (Irish)	0.03	81	3.0	82.8	1.7
Maize Flour Loose	0.03	264	2.4	299.6	1.5
Fresh fish	0.02	121	22.7	12.3	1.3
Rice Grade 1	0.02	346	8.4	85.1	1.1
Onion & Leeks	0.02	44	5.0	16.9	1.1
Eggs	0.02	152	10.6	25.3	1.0
Cabbages	0.02	21	2.0	18.1	1.0
Chicken	0.02	163	2.0	130.3	0.9
Banana Ripe	0.02	67	3.5	30.5	0.9
Tea	0.02	40	33.4	1.9	0.9
Cooking Oil	0.02	900	20.5	66.9	0.8
Oranges	0.01	43	4.2	12.8	0.7
Cooking Banana	0.01	64	3.2	24.9	0.7
Wheat Buns	0.01	370	10.0	44.5	0.7
Avocado	0.01	128	3.0	47.1	0.6
Mutton & Goat	0.01	166	21.8	8.2	0.6
Offal (Matumbo)	0.01	143	10.5	14.4	0.6
Margarine	0.01	745	20.5	38.0	0.6
Dried & Smoked	0.01	269	28.0	9.5	0.5
Mangoes	0.01	43	3.8	11.6	0.6
Green Maize	0.01	56	2.5	20.2	0.5
Millet	0.01	318	4.5	54.0	0.4
Carrots	0.01	38	3.0	9.1	0.4
Beef without Bones	0.01	195	23.3	5.9	0.4
Pasta (Spaghetti)	0.01	111	10.3	7.2	0.4
Wheat Flour	0.01	320	5.6	36.9	0.4
Spinach	0.01	32	4.2	4.7	0.3
Grams	0.01	322	7.0	25.7	0.3
	1.00	8750.0	360.3	4,094.3	55.1
Monthly per adult equivalent					1,652.1

Appendix 2: Poverty Derivation for Mar. 2008

Food item	Share in basket	Kcal (100g)	CPI price (KShs/100g)	Kcal per KShs 100	KShs Per 2,250 Kcal
Beef with Bones	0.09	156	19.7	71.3	5.5
Sugar	0.08	375	7.4	396.6	4.8
Milk-Fresh Unpacked	0.07	72	6.5	75.1	4.2
Bread	0.07	257	8.3	204.6	4.1
Maize Flour Sifted	0.06	264	2.8	557.7	3.6
Milk (Fresh Packed)	0.05	72	6.5	50.3	2.8
Tomatoes	0.04	27	4.7	24.7	2.6
Rice (Grade 2)	0.04	346	5.6	247.6	2.4
Maize Grain Loose	0.04	353	2.2	587.8	2.2
Kale (Sukuma Wiki)	0.04	52	3.1	59.8	2.2
Beans	0.03	314	3.9	265.3	2.0
Cooking Fat	0.03	900	19.6	147.4	2.0
Potatoes (Irish)	0.03	81	3.7	66.2	1.9
Maize Flour Loose	0.03	264	2.8	256.2	1.7
Fresh fish	0.02	121	25.8	10.9	1.4
Rice Grade 1	0.02	346	8.9	80.1	1.3
Onion & Leeks	0.02	44	5.9	14.2	1.2
Eggs	0.02	152	11.2	23.8	1.1
Cabbages	0.02	21	2.7	13.3	1.1
Chicken	0.02	163	2.7	96.2	1.0
Banana Ripe	0.02	67	3.9	27.2	1.0
Tea	0.02	40	36.8	1.7	1.0
Cooking Oil	0.02	900	24.9	55.2	0.9
Oranges	0.01	43	5.0	10.8	0.8
Cooking Banana	0.01	64	3.6	22.2	0.8
Wheat Buns	0.01	370	10.3	43.0	0.7
Avocado	0.01	128	3.1	45.3	0.7
Mutton & Goat	0.01	166	23.1	7.7	0.7
Offal (Matumbo)	0.01	143	11.3	13.4	0.6
Margarine	0.01	745	21.1	36.9	0.6
Dried & Smoked	0.01	269	30.6	8.7	0.6
Mangoes	0.01	43	4.1	10.6	0.6
Green Maize	0.01	56	2.9	17.6	0.6
Millet	0.01	318	4.8	50.6	0.5
Carrots	0.01	38	4.0	6.8	0.4
Beef without Bones	0.01	195	24.7	5.5	0.4
Pasta (Spaghetti)	0.01	111	11.2	6.6	0.4
Wheat Flour	0.01	320	6.0	34.3	0.4
Spinach	0.01	32	4.5	4.4	0.4
Grams	0.01	322	7.9	22.9	0.3
Total	1.00	8750.00	398.1	3,680.4	61.3
Monthly per adult equivalent					1,837.8

Appendix 3: Poverty Derivation for Apr. 2008

Food item	Share in basket	Kcal (100g)	Median rural price (KShs/100g)	Kcal per KShs 100	KShs Per 2,250 Kcal
Beef with Bones	0.09	156	19.4	72.5	5.9
Sugar	0.08	375	7.1	415.7	5.2
Milk-fresh unpacked	0.07	72	6.5	75.5	4.5
Bread	0.07	257	8.4	202.8	4.4
Maize Flour Sifted	0.06	264	3.2	498.5	3.9
Milk –Fresh Packed	0.05	72	6.5	50.6	3.0
Tomatoes	0.04	27	6.6	17.6	2.8
Rice (Grade 2)	0.04	346	6.0	228.7	2.6
Maize Grain Loose	0.04	353	2.4	536.8	2.4
Kale-Sukuma Wiki	0.04	52	3.0	61.3	2.3
Beans	0.03	314	6.7	155.5	2.2
Cooking Fat	0.03	900	20.0	144.3	2.1
Potatoes (Irish)	0.03	81	3.8	65.6	2.0
Maize Flour Loose	0.03	264	3.2	229.0	1.8
Fresh fish	0.02	121	24.2	11.6	1.5
Rice Grade 1	0.02	346	9.2	77.7	1.4
Onion & Leeks	0.02	44	6.3	13.3	1.3
Eggs	0.02	152	11.2	23.8	1.2
Cabbages	0.02	21	2.5	14.3	1.1
Chicken	0.02	163	2.5	103.2	1.1
Banana Ripe	0.02	67	3.8	27.9	1.0
Tea	0.02	40	36.0	1.7	1.0
Cooking Oil	0.02	900	25.7	53.5	1.0
Oranges	0.01	43	4.5	12.0	0.8
Cooking Banana	0.01	64	3.5	22.6	0.8
Wheat Buns	0.01	370	10.8	41.2	0.8
Avocado	0.01	128	3.4	41.6	0.7
Mutton & Goat	0.01	166	22.7	7.9	0.7
Offal (Matumbo)	0.01	143	10.9	13.8	0.7
Margarine	0.01	745	21.2	36.7	0.7
Dried & Smoked	0.01	269	29.5	9.0	0.7
Mangoes	0.01	43	3.9	11.3	0.7
Green Maize	0.01	56	2.9	17.7	0.6
Millet	0.01	318	5.4	45.0	0.5
Carrots	0.01	38	5.6	4.9	0.5
Beef without Bones	0.01	195	24.2	5.6	0.5
Pasta (Spaghetti)	0.01	111	11.7	6.3	0.4
Wheat Flour	0.01	320	6.0	34.7	0.4
Spinach	0.01	32	4.9	4.0	0.4
Grams	0.01	322	8.0	22.6	0.4
Total	1.00	8750.00	403.2	3,418.1	66.0
Monthly per adult equivalent					1,978.9

Appendix 4: Poverty Derivation for Feb/Mar. 2010

Food item	Share in basket	Kcal (100g)	Median rural price (KShs/100g)	Kcal per KShs 100	KShs Per 2,250 Kcal
Beef with Bones	0.09	156	23.4	60.0	7.1
Sugar	0.08	375	9.4	313.0	6.2
Milk-fresh unpacked	0.07	72	7.5	65.2	5.4
Bread	0.07	257	8.7	96.8	5.2
Maize Flour Sifted	0.06	264	4.2	374.8	4.7
Milk-fresh Packed	0.05	72	7.5	43.7	3.6
Tomatoes	0.04	27	5.4	21.7	3.4
Rice (Grade 2)	0.04	346	7.1	195.4	3.1
Maize grain Loose	0.04	353	3.4	379.2	2.8
Kale-Sukuma Wiki	0.04	52	3.0	61.4	2.8
Beans	0.03	314	7.6	135.7	2.6
Cooking Fat	0.03	900	17.7	163.7	2.5
Potatoes (Irish)	0.03	81	3.6	69.3	2.4
Maize Flour Loose	0.03	264	4.2	172.2	2.2
Fresh fish	0.02	121	28.2	9.9	1.8
Rice Grade 1	0.02	346	13.3	53.8	1.6
Onion & Leeks	0.02	44	7.2	11.8	1.5
Eggs	0.02	152	11.2	23.8	1.4
Cabbages	0.02	21	2.5	14.5	1.4
Chicken	0.02	163	2.5	104.7	1.3
Banana Ripe	0.02	67	4.9	21.5	1.2
Tea	0.02	40	36.0	1.7	1.2
Cooking Oil	0.02	900	22.9	60.0	1.2
Oranges	0.01	43	5.5	9.8	1.0
Cooking Banana	0.01	64	4.2	18.7	1.0
Wheat Buns	0.01	370	11.0	40.4	0.9
Avocado	0.01	128	3.6	39.7	0.9
Mutton & Goat	0.01	166	27.3	6.5	0.8
Offal (Matumbo)	0.01	143	12.9	11.7	0.8
Margarine	0.01	745	24.5	31.7	0.8
Dried & Smoked	0.01	269	30.6	8.7	0.8
Mangoes	0.01	43	4.8	9.1	0.8
Green Maize	0.01	56	3.5	14.7	0.7
Millet	0.01	318	6.2	39.2	0.6
Carrots	0.01	38	5.3	5.1	0.6
Beef without Bones	0.01	195	30.1	4.5	0.6
Pasta (Spaghetti)	0.01	111	14.3	5.2	0.5
Wheat Flour	0.01	320	5.4	38.7	0.5
Spinach	0.01	32	5.3	3.7	0.5
Grams	0.01	322	10.1	17.8	0.4
Total	1.00	8750.00	445.9	2,859.1	78.9
Monthly per adult equivalent					2,365.8

Chapter 8
Assessing ICT Impact on Development and Economic Transformation in the Rural Areas of Tanzania:
The Case of Rural Tele-centres

Peter Ulanga

1. Introduction

The last quarter of the 20th century has seen tremendous growth in use of ICTs in many aspects of the economy. In Tanzania ICT is the fastest growing sector with the annual growth rate of over 29%. The current penetration of telephony services is around 26% covering most urban centres and with significant penetration in rural areas (TCRA, 2008). Information and Communication Technologies consist of infrastructure used to transmit information, information storage mechanism in computers and servers and the applications that run in them. They consist from the most basic form such as broadcasting and basic telephony to the most advanced systems like grid computing where computers share processing capacity to solve very large scale computational problems. The ICT Sector can be divided into three sub-systems namely technology and infrastructure (comprising communication and data processing), institutions (comprising legal and economic institutions) and social structures and processes. These influence ICT usage and the interpersonal links that facilitate the transfer of information (Wangwe, 2007, 2008). In looking at the three subsystems of ICT sector it can be found that the extent at which the first subsystem is deployed in rural Tanzania is limited. However, the influence of this limited extent of deployment should be known.

Earlier studies have been done in trying to find out the economic impact of telephony on rural livelihoods and poverty reduction. Souter et al. (2005) found consistent behavioural pattern of telephone usage in the three countries namely in Tanzania, India and Mozambique. That is, telephones were considered very important for use in emergencies, extensively used to maintain social networks, especially contact within the family, valued more for saving money than for earning money, valued more by richer and better educated people than by the poorer, less educated or more marginal members of society – especially where financial value was concerned and considered unimportant for information gathering. The work further found out that 45% of the telephones were acquired within one year of the study and 33% of those without phones said they would acquire one within a year after the study. The trend shows a very consistent high demand for telephone services. The study did not cover rural areas with access to internet; hence nothing is said on the use of internet. In 2004 only 10% of the respondents to an interview conducted by Ngalinda and Mutagahywa did own mobile phones (Ngalinda and Mutagahywa, 2005). Out of those 48% lived in Dar es Salaam and only 8.3% lived in rural areas with about 44% living in other urban areas in Tanzania (Ngalinda and Mutagahywa, 2005). The same study found that only 2% of the respondents did have e-mail addresses, most of them living in urban centres. Also it was found that only 2% of the Tanzanian households had at least 1 working computer. Of them, only 15% have access to internet and 94% were in Dar es Salaam (Ngalinda and Mutagahywa, 2005). The dismal picture painted by these statistics show that rural Tanzania is least connected hence the focus on community based tele-centres, where rural based communities can have access to ICTs.

Another developing phenomenon is the presence of General Packet Radio Service (GPRS) enabled Global System for Mobile Communication (GSM) networks and mobile phones which can provide narrow band internet access to mobile phones. While their use in rural setting has not been investigated fully, proliferation of cheap GPRS enabled phones the use of narrow band internet which can grow in rural areas. This study has looked at the use of ICTs in community based telecentres and in areas with access to telephones, computers, internet, radio and television. It did not look at the lack of access, instead, the impact of other dynamics for example level of education, income distribution, type of occupation and age group distribution were investigated in addressing the impact of ICT adoption. The study also looked at how the ICTs were used in four areas namely education, entertainment, economic activities and information sharing and gathering. The findings have shown in aggregate the ICTs are used as information sharing and gathering tool at most and for economic activities in the least.

This chapter is structured in six sections. The first section introduces the issues, the second section consists of the status of the ICTs in Tanzania covering literature review, then the third section presents data and descriptive analysis, the fourth section analyses the results and the fifth section presents key results. Finally the sixth section provides conclusions.

2. Status of ICTs in Tanzania

In looking at the impact of ICTs in development and economic transformation in Tanzania various efforts in terms of policy, legal and regulatory initiatives have been reviewed. The first important document to be reviewed has been the National ICT Policy which was adopted in 2003 to spearhead the ICT development efforts in the country. Tanzania's National ICT Policy is aligned to the following vision statement: "Tanzania to become a hub of ICT infrastructure and ICT solutions that enhance sustainable socio-economic development and accelerated poverty reduction both nationally and globally"(MOCT, 2003:2).

The overall mission of this policy is: "To enhance nation-wide economic growth and social progress by encouraging beneficial ICT activities in all sectors through providing a conducive framework for investments in capacity building and to promote multi-layered co-operation and knowledge sharing locally as well as globally" (MOCT, 2003: 2). Now to phrase the policy context, "The Tanzania development vision 2025 envisages a nation with five main attributes: high quality livelihood; Peace, stability and unity; good governance; a well educated and learning society; and a strong and competitive economy capable of producing sustainable growth and shared benefits" (MOCT, 2003:2). On the other hand this policy document has articulated ten main focus areas in harnessing ICT in Tanzania which include strategic ICT leadership; ICT infrastructure; ICT Industry; Human Capital; Legal and Regulatory Framework; Productive Sectors; Service Sectors; Public Service; Local Content; and Universal Access (MOCT, 2003: 2).

Figure 1: Telephony penetration from year 2000 to June 2008

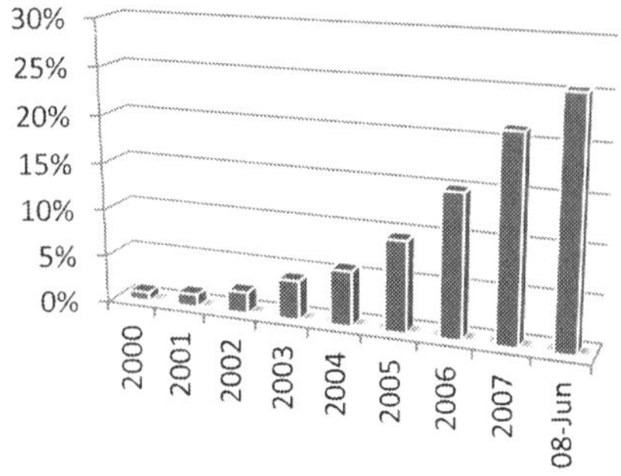

Tanzania is one of the countries with very low GDP in the world and has a high percentage of its population engaged in subsistence farming. Yet in ICT services the country has seen significant growth from 5 operators in 2000 to 10 operators by end of year 2008, and from about 300,000 users of telephones mobile and fixed in 2000 to over 10.4 million users in 2008, with a tele-density growth from around 1% in 2000 to over 25% in year 2008. This shows a potential of opportunities waiting to be explored and perhaps exploited. By the year 2008, the ICT Policy of 2003 and the introduction of the New Licensing Framework which is technology and service neutral has pushed the sector into an unprecedented growth path driven largely by private sector participation in a highly competitive market. ICT now plays a fundamental role in the economic development as the country is evolving into a knowledge-based economy. The government zero-rated taxation on computer hardware to accelerate the use of ICTs. While this effort is lauded as an important step towards increasing penetration of ICT hardware it has affected negatively the ICT hardware production industry where companies which were engaged in assembling of computers could not compete with the imported clone computers. Furthermore, software piracy has been getting out of control. Both these problems have muted the production side of ICTs locally. However, the country is still in track towards the knowledge economy as envisioned in the National ICT Policy adopted in 2003. The National ICT Policy states the role of the Government in ICT sector in relation to its stated objectives. The policy states strategically on how to make ICT as a strategic tool for implementation of the Tanzania Development Vision 2025, which among other things advocate for a new thinking in the country (MOCT, 2003). The Vision 2025 demands the cultivation of a developmental mind set, empowering culture, competence, competitiveness, good governance and the rule of law.

In a Tanzania IT-Vendor Survey conducted by UNDP Tanzania, Dar es Salaam Institute of Technology (DIT) and the Institute of African Affairs (Hamburg) the results showed a high concentration of activities in Dar es Salaam. Also the market is hardware based with limited software development activities. Also a fair distribution of IT-users and utilisation points towards business opportunities in different market segments (Nielinger, 2002:3).

In year 2000, the country had 5 operational licensees with infrastructure based licences. By the end of 2008 the number of licensees has soared to 10 with 6 operational, 2 building infrastructure for a commercial service and two more planning to start building infrastructure. At the same time application service licences have increased from 11 in 2000 to over 42 by the end of year 2008. Between 2000 and 2006, there was also a significant increase in the number of radio (from 14 to 47) and TV stations (10 to 29). However, only five of the radio stations have a national coverage and the majority have district level coverage.

A Miller, Esselaar and Associates (Miller et al., 2001:18) estimates year 2000/1 ICT industry to be about US$ 300 - 350 million with the incumbent operator Tanzania Telecommunication Company Limited (TTCL) generating about US$ 138 million, the high revenue could also be attributed to the market position they had at the time with exclusivity on international calls. The overall growth rate is in excess of 25% per annum fuelled by buoyant sales in the cellular industry (Miller et al., 2001:18).

As per the Tanzania National ICT Policy of 2003 (MOCT, 2003) very few educational institutions have computer laboratories and other multimedia facilities and these are more prevalent in private educational institutions than in public ones. In any event facilities are insufficient to meet demand. In addition while there is an official Secondary School Computer Studies Syllabus for Forms I – IV, developed in 1996 and issued in 1997, it is out of date with respect to the evolution of technology since the early 1990s. Furthermore, the lack of a programme for training teachers on computers and other multi-media utilization has been identified as a major reason for slow take up of computer studies in primary and secondary schools. It was suggested that typing skills and the use of "mockups" as teaching aids to simulate computers and peripherals should be promoted in schools that cannot afford to purchase ICT equipment (MOCT, 2003). In general, there is a shortage of well-qualified professionals of ICT in Tanzania. There are also no well-established ICT professional profiles, and a standardised process of evaluation or certification of the different courses offered by various training centres is lacking. Access to online and distance learning for ICT is also still limited. Furthermore, opportunities for training are mostly limited to few urban centres. While there are many Tanzanian websites, most are in English, and are not, therefore, a dominant medium for society to access information. Many websites are not updated regularly and appear to be an advertising presence on the Web. However some are vibrant websites with the majority publishing local news on the Web, while others demonstrate some convergence by giving access to local radio programmes on the Internet. There has been recent controversy on access to pornography via the Internet causing concern for safeguarding of the diverse mores, morals and culture. The potential for e-commerce is constrained by the lack of local credit cards and an appropriate legal framework that engenders an environment of trust, security and accountability.

According to Kijo-Ringo (2002), who covered 450 SMEs, 150 from each of the three EA economies, Tanzania SMEs play a dynamic role in promoting economic growth and development by contributing to increased employment and creation of income opportunities. The SME sector is believed to contribute about 1/3 of Tanzania's gross domestic product, according to 2002 Economic Survey (Kijo-Ringo, 2002). The survey investigated the link between the investments in and utilization of ICTs by Tanzanian SMEs and market. The methodology used involved analysing the performance of such SMEs, by looking at their geographical extension through the use of micro level data of the East Africa SMEs survey. Area sampling was used whereby, 118 samples of SMEs from major commercial corridors in the country (Lake Zone, Coastal Zone and Arusha region) were randomly selected and distributed equally in the food processing, textile and tourist sectors. The two key considerations in the determination of the sample regions were their economic significance and their ability to represent fairly the SMEs sector. In the study it was found that majority of the sectors used public telephones but in the case of computer based ICTs (internet, fax and e-mail), it was mostly used in the tourism sector and much less in the other sectors. Investment in ICTs by SMEs in the tourism sector was around 60% while for other sectors it was very low around 8.9% and 18.4% for textile and food processing respectively. Considering that investment in ICTs by SMEs leads to increased access to business information such as end-product markets and thereby increased market outreach and extension, it is clear that most of the SMEs are on the disadvantage side (Kijo-Ringo, 2002).

Another important observation captured from this study was that enterprises which utilized ICTs and especially relatively modern and cheap technology like mobile phones are the ones that got access to market information and thereby increased their chances to expand their markets regionally (most of these were from the tourism sector). The study made the following recommendations: (i) SMEs should be advised accordingly on making relevant and appropriate investments in ICTs; (ii) Educate and encourage SMEs operators on the benefits of utilizing ICT facilities such as low transaction cost, support formulation of business networks and reduction of market uncertainties; (iii) Raise awareness among SMEs operators on the importance of maintenance on ICTs facilities so that they function properly for a long time; (iv) SMEs should be advised and encouraged to form groups to facilitate the channelling of ICT related resources and efforts; and (v) Support schemes should be established to facilitate SMEs' use of ICTs by financing their concerns such as investments/utilization, consultation, training fees, repair and maintenance service charges.

Mahegere (2008) argue that ICTs are a powerful tool for development and poverty reduction, if ICT policies together with appropriate legal, regulatory and institutional framework facilitate wider use and accessibility of these technologies to the majority. He also stated that the degree of poverty reduction depends on higher growth in the rural economy, particularly in the agricultural sector. Therefore, he concluded that as far as the overall economic development of Tanzania is concerned, rural areas are priorities (Mahegere, 2008). Heeks (2005) emphasized that, 'all that ICTs do is handle information'. Therefore, the contribution of ICTs to socio-economic development must be found by understanding the role of information in development and access to information and its impact on socio economic development.

ICTs re-enforces the role of information in development and plays a prominent role in the broader conception of poverty by providing faster, efficient and reliable means for creating, storing and disseminating information. Hence, ICTs provide important tools for improvement of health, education, governance, and enhance people's capabilities. ICT offer new channels for the diffusion of knowledge and create physical and virtual spaces for social communication. To be effective, regulators in developing countries need to understand the needs of the poor, their location and the real barriers to their access to adequate services. This arises due to the concern that regulators receive most of their information from the regulated firms, politicians and higher income groups or elites (Parker, Kirkpatrick and Theodorakopoulou, 2005). Muriethi (2002) also reported that in the case of Tanzania, competition and liberalization of the telecommunication sector have shifted the focus of operators to the profitable urban areas at the expense of the rural areas despite the main policy emphasis on improving access in rural areas. Also failures of the operators to reach their universal service targets hinder effective utilization of the ICTs services for poverty reduction especially in rural areas of Tanzania, The Government established a Universal Communications Access Fund to address the disparity. The World Economic Forum has ranked Tanzania in the 11th position in Africa for information and communication technology strength. The forum's Global Information Technology Report 2006-2007, Networked Readiness Index, had ranked Tanzania as the 91st in the worldwide. The index compared the status quo of the indexed countries and regions with their strengths and weaknesses two years ago. But the forum warned that sub-Saharan African countries had almost all dropped from their previous rankings except Nigeria. Tanzania was 83rd in the 2004-2005 index, but dropped to 84th in the 2005-2006 index.

The World Economic Forum report identified some factors that were hindering the ICT improvement and development in the sub-Saharan region. These included: lack of extensive and efficient infrastructure, over-regulated business environment, and low education standards. For providing all inclusive input to policy development process for ICT, Tanzania Ethink Tank was formed. The eThinkTankTz's vision was stated thus: "to offer ICT leadership by catalysing policy changes and by supporting related developments aimed at enabling Tanzanians to participate effectively in the modern Internet based global economy, benefiting their nation and partners". The mission was articulated as: "to a) catalyse the formation of a national ICT organisation, with mandates from the Government and b) support the national ICT organisation" (Ethink Tank, 2002: 1). It is from eThinkTank's initiative the National ICT Policy of 2003 was formulated.

3. Policies affecting ICTs in rural areas

In order to investigate the impact of ICTs in rural transformation and development of socio-economic systems in rural areas it is important to revisit various policy statements which were directed to ensure development in rural areas. Four policies and one strategy will be discussed here namely Vision 2025, National Telecommunication Policy, National Broadcasting policy, National ICT Policy and Rural Development Strategy. Vision 2025 has five main attributes namely high quality livelihood; peace, stability and unity; good governance; a well educated and learning society; and a competitive economy capable of producing sustainable growth and shared benefits. Focusing on these five major attributes it can be seen that ICTs has a role as a catalyst or an enabler to facilitate access to information and making communication within the society easy and between society and its leadership. The vision 2025 has made specific statements related to development of ICTs (Planning Commission, 2000: 21): (iv) Promotion of Information and Communication Technologies (ICTs)... Advanced micro-electronic information and communication technologies (ICTs) are central to a competitive social and economic transformation. ICT costs are continuing to fall while their capabilities and resultant profitability enhancements are increasing. These technologies are a major driving force for the realization of the Vision. They should be harnessed persistently in all sectors of the economy and should be put to the benefit of all social groups with a view to enabling the meeting of basic needs of the people, increasing productivity and promoting competitiveness.

The expectations are that the use of ICTs would among other things increase productivity and promote competitiveness in our economy. This has thus introduced a dimension of making ICTs as central in ensuring competitive social and economic transformation. With the decreasing costs of ICTs and proliferation of ICT related services, ICTs could play an increasing role in the development and economic transformation in Tanzania. National Telecommunication Policy (NTP) aims at ensuring the accelerated development of an efficient telecommunications network that can provide an info-communication infrastructure and universal access to telecommunications services for all sectors of the national economy and segments of the population. This policy was set up in 1997 when the infrastructure was predominantly fixed. The target set for tele-density was 6% by year 2020, currently the tele-density is over 26%. Revisiting the National ICT Policy, which is the most current policy document, related to ICTs, the policy envisioned "Tanzania to become a hub of ICT Infrastructure and ICT solutions that enhance sustainable socio-economic development and accelerated poverty reduction both nationally and globally." Also it has the following mission "To enhance nation-wide economic growth and social progress by encouraging beneficial ICT activities in all sectors through providing a conducive framework for investments in capacity building and in promoting multi-layered co-operation and knowledge sharing locally as well as globally." Specifically the policy did want to ensure that the national ICT policy does not exacerbate the digital divide among the people of Tanzania; it must contain provision for bringing access to the more remote areas of the country and those under served in urban areas. Another important document is the rural development strategy launched in 2001, which among others was stipulating a holistic approach towards rural development. The approach has 6 major blocks namely increased opportunities and access to services, promoted widely shared growth, reduced risks and vulnerabilities, production, leadership and governance, and infrastructure and communication. Rural Development Strategy Framework has identified telecommunications in the group of infrastructure and communication as one of the key inputs to rural development. It also provides increased opportunities and access to services. Interestingly two important inputs have been identified namely private sector participation and community management. In the sector development objectives it is envisaged to provide adequate, sustainable and efficient telecommunication services in all sectors of the economy with interconnectivity internationally. With the ICT policy in place the implementation of it has faced many challenges. The growth which is seen is mainly organic and not coordinated. The supply of IT professionals is considerably less than current demand, especially in the areas of higher skills and experience. Furthermore, job mobility in the ICT sector is very high. Only few local websites recently began offering limited e-business services. However these services are constrained by the lack of a national payment system, local credit cards, and a legislative framework appropriate for e-business.

Previous Research Work

- The Economic Impact of Telecommunications on Rural Livelihoods and Poverty Reduction: a study of rural communities in India (Gujarat), Mozambique and Tanzania. -2005 (Souter et al., June 2005).
- Towards an African e-Index: ICT access and usage (Ngalinda and Mutagahywa, 2005).
- Impact of Investment in and Utilization of ICTs on market extension (Kijo-Ringo, 2002).

4. Data and Descriptive Analysis

As part of the major study on assessing ICT impact on development and economic transformation in Africa, this study covered 750 respondents in 17 localities in the rural areas of Tanzania. It looked at the use of five forms of ICTs namely the radio, telephone, internet, computer and television. Out of these the radio was the most widely used with 719 respondents using it followed by the telephone with 672 respondents, closely followed by the television with 659 respondents. The internet and computer was being used by 459 and 455 respondents respectively. This can be attributed to the level of skills required in the use of internet and computers compared to the use of radio, telephone and television. The cost could be the main determinant behind the use of radio telephone and television. There is a small variation of users across different income brackets for radios (7%) with a bigger variation across income brackets for telephone and television (15% and 12% respectively). There is a large variation for computer users and internet users across education distribution. For example, out of 84 respondents who have education qualification less than primary education (though not completely illiterate) only 11% were using the internet and 13% were using the computer, while 81% were using radio and 54% were using television, and 49% were using the telephone. At the level of standard 7 after completion of primary education there is some improvement with 34% using the internet while radio, television and telephone are used by 95%, 82% and 86% respectively. The differences seem to converge with increased level of education at the degree or diploma level the percentages are comparable with a difference of less than 10%.

The study looked at four main uses of ICTs namely for education, entertainment, economic activities and information. The aggregate of the 5 forms of ICTs investigated shows that in the areas sampled the most common use is for information sharing followed by entertainment then education and finally economic activities (96%, 94%, 84% and 75% respectively). The respondents used radio primarily for information and entertainment (88% of respondents), while 77% used the radio for education and 60% to obtain information related to economic activities, making economic activities the last priority in the use of radio though it is still significant.

About 87% of the respondents used telephone mainly for information sharing and 67% used it for economic activities. Due to the nature of telephone services almost no use for education or entertainment. On the other hand the internet was mainly used for information gathering and education purposes (56% and 55% respectively) followed by the use for entertainment (49%), still economic activities lagged behind at 46%. The television on the other hand had information and entertainment as the leading uses (80% and 79% respectively) compared to 70% of users who used it for education and only 55% used it for economic activities. The computer has been used mainly for education purposes followed by information gathering then entertainment followed by economic activities. This study looked at 17 telecentres. Out of the 17 telecentres each one had between 2 to 36 computers, 7 out of the 17 telecentres did not have telephone lines. All except one had at least one of these additional facilities namely printers, photocopiers or facsimile machines. 11 out of 17 had at least one of these services namely sale of recharge vouchers (airtime), charging of cell phones, sell of postcards and stamps, document scanning services, document binding services, lamination, printing, internet, computer training and farmers training. The age of the computers varied from slightly over 1 year to 10 years with some computers still using Windows 98 as the operating system with Windows XP as the most predominant operating system. The level of education of the telecentre operators varied from primary education to advanced secondary education.

5. Analysis and Key Findings

Table 1: Use of ICTs according the Economic Activities

	Unemployed	Traders	Employed	Farmers
Radio	95%	97%	100%	90%
Telephone	90%	96%	96%	70%
Internet	63%	62%	73%	32%
Television	90%	90%	95%	68%
Computer	66%	63%	70%	31%

The respondents were divided into four groups un-employed, traders, employed and farmers. The patterns in the use of ICTs were investigated. In total out of the 750 respondents 278 were unemployed, 156 were traders, 201 were employed and 115 were farmers. In all four categories at least 90% of the respondents used radios. About 90% of farmers, 97% of the traders, 95% of the unemployed, and all employed respondents used radio. The telephone was used most pre-dominantly by the traders and employed respondents at 96% followed by the unemployed at 90% and finally farmers at 70%. The use of internet was most predominant in the employed group of the respondents at 73%, followed by unemployed and traders at 63% and 62% respectively and the trailing group was farmers at only 32%. The use of television was most pre-dominant in the employed group at 95%, followed by unemployed and traders at 90% and farmers at 68%. The use of computer followed a similar pattern to the use of internet. The leading users being employed at 70%, followed by the unemployed at 66%, traders at 63%, and finally the farmers at 31%.

Figure 2: Use of ICTs according to the income distribution and type of economic activity ('000 Tshs)

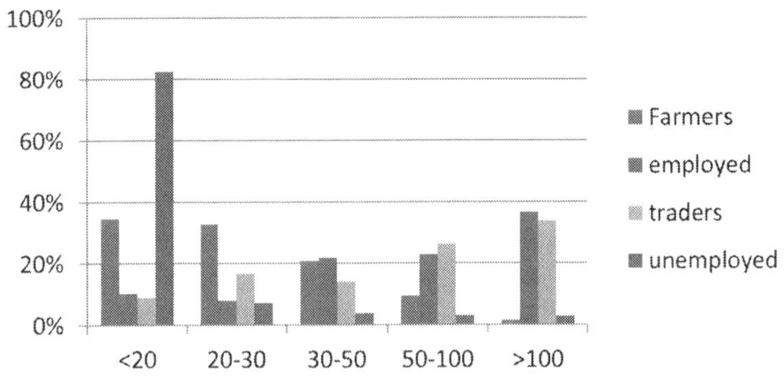

Figure 3: Use of ICTs according to the levels of Education and type of economic activity

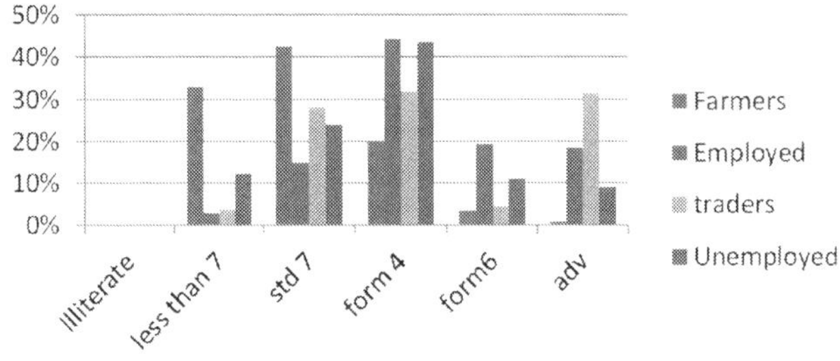

While income distribution could explain the higher use of ICTs to traders and employed it could not explain the high use of ICTs among the unemployed who largely belong to the lowest income bracket (see Figure 2). The unemployed have a different education profile compared to farmers with majority of them at form 4 level (the same level as employed), while majority of farmers are at standard 7 and less than standard 7. This could be a significant factor in the use of ICTs especially the internet and computers as shown in the previous analysis (see Figure 3). This can be summarised by the figure below looking at the percentage use of all 5 different types of ICTs across different education groupings showing that use of computers and internet is much lower for standard seven leavers compared to form four leavers (see Figure 4).

In aggregate over 75% of the users use ICTs daily (see Figure 5) with very few using ICTs once per year. A breakdown of use of ICTs according to the type of economic activities has shown some differences especially with farmers (see Figure 6). While daily use was more prevalent for radio, telephone and computers, for internet weekly use was more prevalent. However this trend was not visible for employee and traders (see Figure 7).

Figure 4: percentage of users of ICTs against levels of education (aggregate)

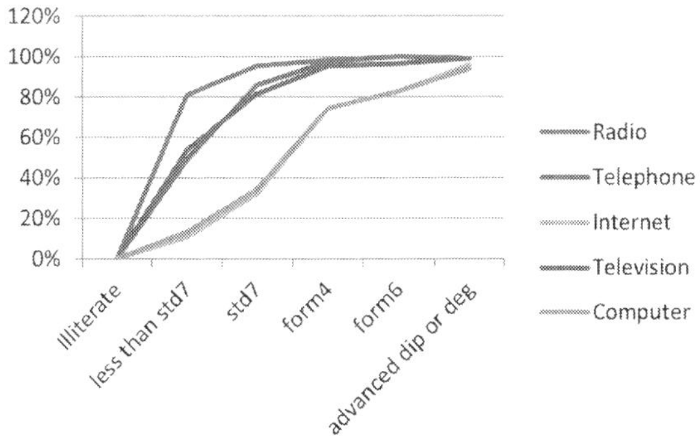

Figure 5: Frequency of use of ICTs according to type of economic activity (aggregate)

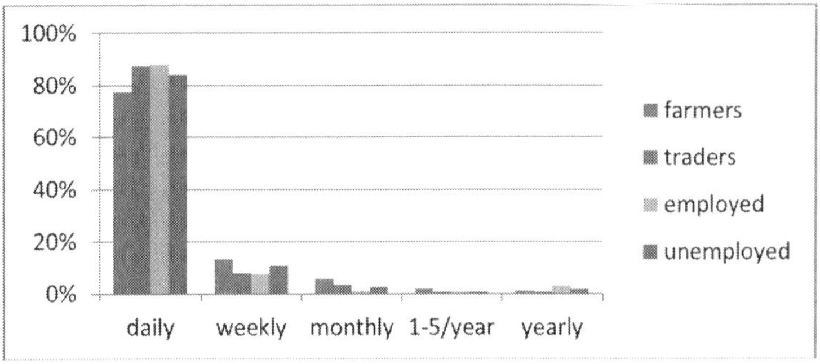

Figure 6: Frequency of use of ICTs by farmers according type of medium

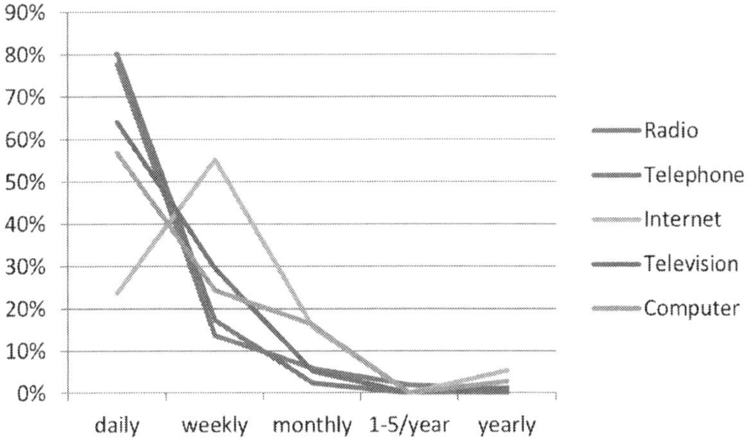

Figure 7: Frequency of use of ICTs by traders according to type of medium

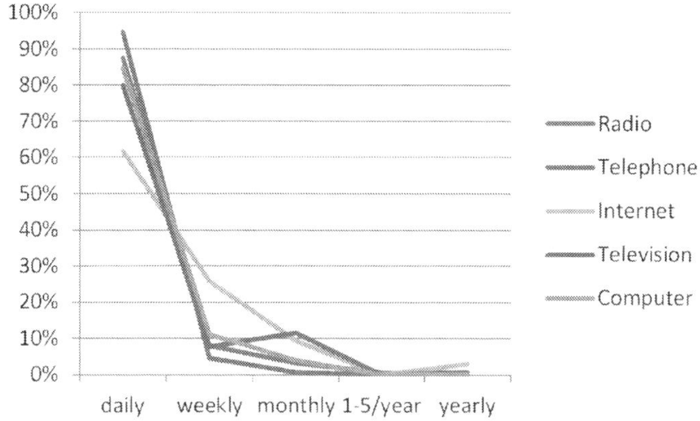

In use of ICTs for economic activities is analysed. It shows that farmers, traders and employed respondents used the telephone most for economy activities, followed by radio and television, and internet and computer were trailing behind.

Figure 8: Percentage of users using different types of ICTs for economic activities

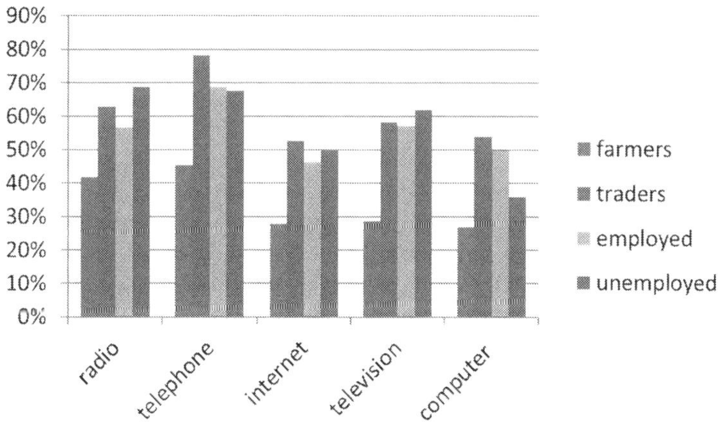

Figure 9: Percentage of users using different types of ICTs for economic activities (primary education certificate)

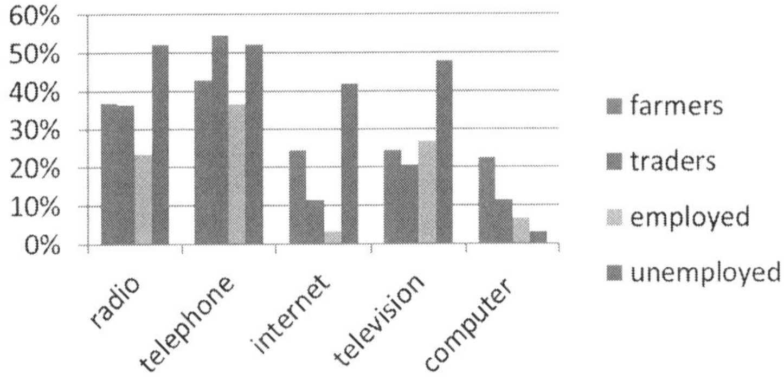

Figure 10: Percentage of users using different types of ICTs for economic activities (form 4 leavers)

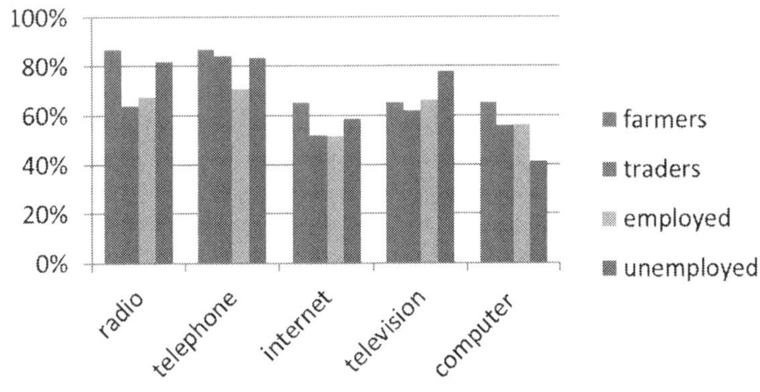

Figure 9 and Figure 10 make an interesting comparison on the use of ICTs for different types of economic activities by taking education level of respondents as a factor, i.e. those who have completed primary education and those completed secondary education. It can be seen clearly that ICT usage for economic activities for those who have completed secondary education is much higher than those who have completed primary education. This difference covers all types of ICTs and all types of trades.

Figure 11: Percentage of users using different types of ICTs for information sharing

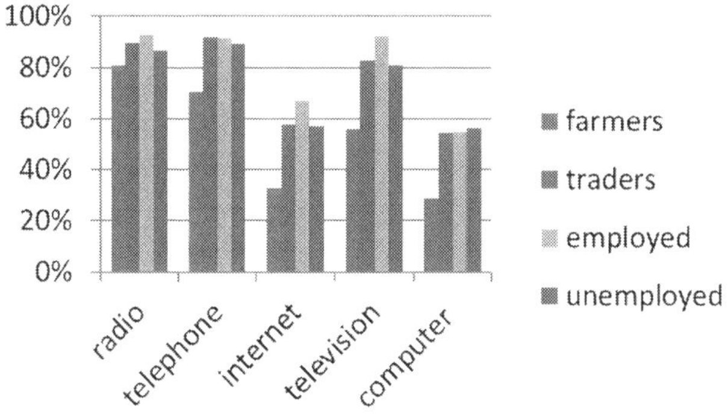

Use of ICTs for information (see Figure 11) in aggregate shows similar trends for radio and telephone and television being the most widely used, with internet and computer trailing behind.

Figure 12 shows a comparison of ICTs used by farmers for economic activities and information services. It shows by order of importance farmers use the radio, telephone and television for information services. They mostly use the telephone and radio for economic activities followed by television, and internet and computer use is trailing behind. The trend is different for traders who use telephone, radio and television for information (see Figure 13) and use of internet and computer is trailing. Likewise, for economic activities they use telephone followed by radio, television, internet and computers in this order.

The use of ICTs in rural areas has been investigated looking at 5 types of media namely radio, telephone, internet, television and computer. The study looked at the 3 main economic activities namely trading farming, formal employment and a group of unemployed people mainly students and school leavers without formal employment. Even in the presence of 5 types of access the preference is still for the use of radios. The reason appears to be the cost, as the radio is the cheapest medium and it is also the easiest to use. Telephone is the second most preferred means of communication. This could be attributed to the fact that it is person to person and hence a more direct means of communication. Although internet also can achieve the same, it is more expensive compared to telephone. Use of ICTs also depends on level of education especially for the use of internet and computers (see Figure 4). It is interesting to note however that there was no trend when the use was compared to income levels, suggesting that

ICTs are considered essential and not luxury or expensive a products at the individual level.

Figure 12: Comparison of use of ICTs by farmers for information and economic activities

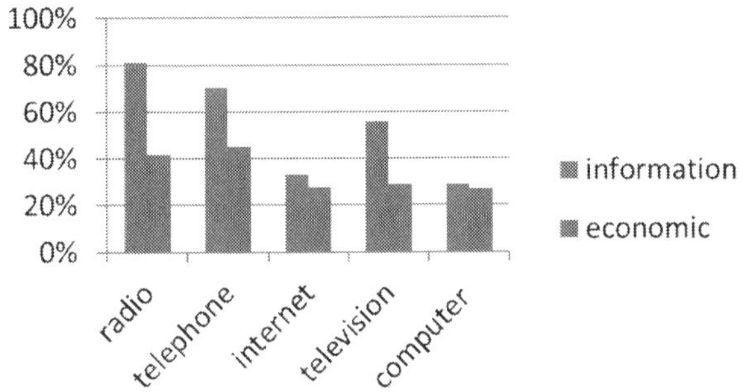

Figure 13: Comparison of use of ICTs by traders for information and economic activities

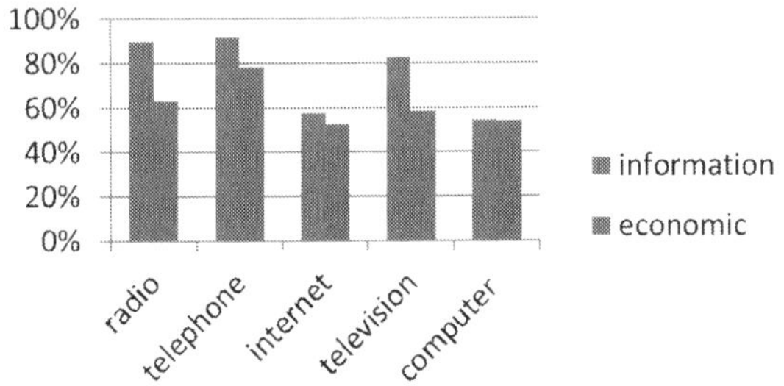

Figure 14: Comparison by employed respondents on the use of ICTS for information and economic activities

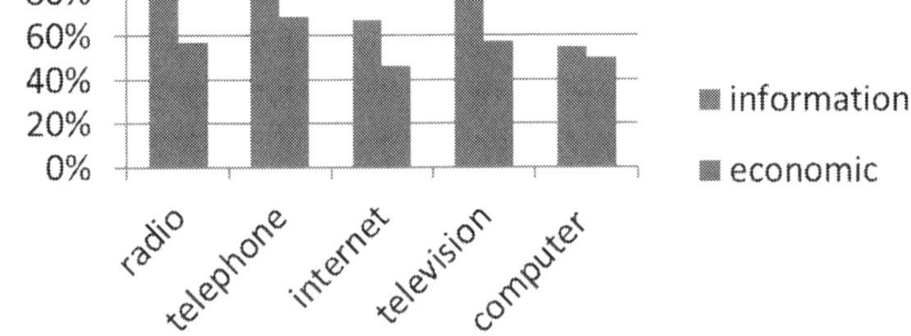

The telecentres act as primary providers of computer facilities and internet facilities on the one hand and on other hand they are secondary providers of telephony services. 7 out of 17 telecentres are not having any telephone facilities and only 4 out of 17 telecentres have more than 1 telephone user per day. If this is compared with the usage of telephones among the 750 respondents, it shows that many users own their phones or have another access to telephone services. Considering number of users visiting the telecentres per day it shows that utilisation of the computers is low. The average usage of computers is 10 hour per day and only 3 telecentres have average utilisation of 50% or more. This can be compared to the fact that less than 30% of the farmers who used computers have used it for information or economic activities and for traders likewise about 50% of traders who have used the computer have used it for information or economic activities. This brings the issue of relevance of the telecentres in rural setting. So far, the telecentres are providing traditional services not earmarked for specific environment where they exist. Out of the 17 telecentres only one is providing farmer training. In terms of revenues 9 out of 17 telecentres have monthly revenues above 200,000 Tanzanian Shillings (approx 150 USD). The telecentres receiving less than 150 USD will not be sustainable over a long term. To complement the income and services offered most of the telecentres have other facilities namely printers, photocopiers and facsimile machines. 10 out of the 17 have all the three facilities, while 4 have only printers and 2 have only printers and photocopiers and one telecentre does not have any of the facilities. 11 out of the 17 telecentres also have other facilities.

6. Key Results

The results of the study show that there is a large push towards extending ICT access in rural areas in Tanzania. Similar to earlier studies (Souter et al., 2005), information sharing has been the most significant use of ICTs. Level of adoption of ICTs has been affected by level of education. That is, persons with less advanced education use less complex forms of ICTs. Computers and internet are favoured by those with at least secondary level education and above. For those using the ICTs income was not a significant factor determining the use of ICTs. This suggests that ICT products are considered as necessity in the daily livelihood and hence users are making rational economic decisions to spend on ICTs as a medium of communication and information. It can be deduced that ICTs are used primarily as means of sharing information and because their use covers different income brackets and different education profiles, they contribute to improvement of social economic development in the rural societies. There is a significant use of

ICTs directly for economic activities in the rural society which depends on the level of education of the users of ICTs. Hence a much bigger impact of use of ICTs in rural areas is experienced on the users who are more educated compared to less educated (based on comparison between primary education leavers and secondary education leavers). Telecentres are primary providers of complex ICT facilities and services (computers and internet), however they have been found to be marginally sustainable, and not providing types of services that will improve utilisation. One area of further study would be to find out what services, if coupled in the telecentres, would spark or improve utilisation of the telecentres.

7. Conclusions

There is a large push towards extending ICT access in rural areas of Tanzania. This study set out to evaluate the adoption of ICTs in these rural areas. Currently, radio and television still remain the most prevalent means to reach out to population for information purposes. For those using the ICTs, income was not a significant factor determining the use of ICTs suggesting that ICTs products are seen as necessity in the daily livelihood and hence users are making rational economic decisions to spend on ICTs as a medium of communication and information. However, it appears that currently the use of ICTs impose high cost on the rural users. Also, it is evident that ICTs are used primarily as means of sharing information by users from different income brackets and different education profiles. It is also evident, the use of ICT largely contribute towards improvement of social economic development in the rural societie in Tanzania. There is a significant use of ICTs directly for economic activities in the rural society which depends on the level of education of the users of ICTs. Hence a much bigger impact of use of ICTs in rural areas is experienced on the users who are more educated compared to less educated (comparison was based on primary education leavers and secondary education leavers). The Tanzanian Government has zero-rated taxes on computers hence making computers cheaper. The effect is still not so significant as other costs incurred by the telecentres like the cost of bandwidth are still very significant in the total cost of operation of telecentres.

Based on the above findings, th study makes following policy recommendations and suggestions for future research: (i) Increasing ICT education in primary school can reduce or even bridge the difference between secondary education users and primary education users and enable use of ICTs by the primary school leavers; (ii) Initiating programmes to increasse ICT literacy among general population is important to spin out new uses of ICTs and make services like e-government services more relevant; (iii)

The economic linkages should be realigned to increase the value of ICTs in economic activities in order to increase effectiveness of use of ICTs in rural areas. For example this can be done by establishing portals which can improve trade for both farmers and traders; (iv) Radio and television should be used more effectively (as these are still popular among rural population) to instill basic forms of ICT education and complement use of internet and computers in reaching out to the population; (v) A clear policy to locate telecentres in every ward or every primary school will enable proliferation of ICT users in the country with a clear benefit of using ICTs to reach out general population; and (vi) Mechanisms should be developed to reduce the cost of bandwidth for the telecentres.

Telecentres are primary providers of complex ICT facilities and services (computers and internet), however they have been found to be marginally sustainable, and not providing types of services that will improve utilisation. One area of further study would to be to find out what services if coupled in the telecentres would spark or increase utilisation of the telecentres.

References

Ethink Tank (2002), 'EthinkTank Vision', Available at: www.ethinktanktz.org.

Heeks, R. (2005), 'Foundations of ICTs in Development,' Available at: www.manchester.ac.uk/idpm/dig/briefings.htm

Kijo-Ringo, N. (2002), 'Impact of Investment in and Utilization of ICTs on market extension', Available at: www.tzonline.org/pdf/impactofinvestmentinandutilization.pdf (Accessed: 29 December 2008).

Mahegere, J. (2008), 'Information and Communication Technologies for Development and Poverty Reduction', in Tanzania: Policy and Regulatory Issues, *Paper presented at the 3rd National Scientific, Technological and Innovation Conference and Exhibitions*, 5- 7 November, Arusha International Conference Centre, Arusha, Tanzania.

Ministry of Communications and Transport (MOCT) (2003), *National ICT Policy*, March. Available at: Ethink Tank Tanzania - www.ethinktanktz.org (Accessed: 29 December 2008).

Miller, J., Esselaar, P., Hesselmark, O. and James, T. (2001), *A Country ICT Survey for Tanzania: Final Report*, Miller Esselaar and Associates (Wilderness, South Africa) prepared for Sida, Stockholm Available at : http://www.tzonline.org/pdf/Tanzania%20ICT%20Survey.pdf

Mureithi, M. (2002), 'Evolution of telecommunication policy reforms in East Africa: Setting new policy strategies to anchor benefits of policy reforms', Available at: www.wits.ac.za/journal/j0301-muriuki-fin.pdf

Nielinger, O. (2002),'Fact sheet : Tanzania IT – Vendor Survery' part of ongoing research 'ICT for Development: The Case of Tanzania' Hamburg: Institute of African Affairs Available at : http://www.ourtanzania.com/IT%20vendors%20 tanzania.pdf

Ngalinda, I. and Mutagahywa, B. (2005), 'Tanzania,' in Gillwald, A. (ed.), *Towards an African e-Index: ICT Access and Usage in 10 African Countries*, Johannesburg: The Link Centre, Wits University.

Parker, D., Kirkpatrick, C. and Theodorakopoulou, F. C. (2005), 'Infrastructure Regulation and Poverty Reduction in Developing countries: A review of the evidence and research agenda,' Available at: www.competitionregulation.org.uk/conferences/Brazil/Papers/Parker_Kirkpatrick_Figueira.pdf

Planning Commission, Tanzania (2000), *Tanzania Development Vision 2025*, Dar-es-Salaam: Government Press.

Souter, D., Garforth, C., Jain, R., Mascarenhas, O., McKemey, K. and Scott, N. (2005), *The Economic Impact of Telecommunications on Rural Livelihoods and Poverty Reduction: a study of rural communities in India (Gujarat), Mozambique and Tanzania*, June, London: Commonwealth Telecommunications Organisation for the UK Department of International Development.

TCRA (2008), Tanzania Communications Regulatory Authority, Available at: www.tcra.go.tz (Accessed: 29 December 2008).

Wangwe, S. (2007), *A Review of Methodology for Assessing ICT Impact on Development and Economic Transformation*, Dar es Salaam: Daima Associates.

Wangwe, S. (2008), *Evolution, Status and Impact of ICT on Economic Development and Transformation in Africa: An Overview,* November, Dar es Salaam: AERC.

List of Acronyms

GPRS General Packet Radio Service, a data bearer in a GSM cellular phone system,

GSM Global System for Mobile Communication,

MOCT the then Ministry of Communications and Transport currently the ICT functions are handled under Ministry of Communications Science and Technology,

NTP Tanzanian National Telecommunications Policy,

TCRA Tanzania Communications Regulatory Authority.

Chapter 9
The Impact of the Nigerian ICT Policy on the Printing and Publishing Industry

John O. Adeoti, Foluso M. Adeyinka & Moses O. Ubaru

1. Introduction

Countries of Sub-Saharan Africa have started to appreciate the fact that the new information economy can present an opportunity to leapfrog to reduce the technological gap between the developed and less developed economies (Perez and Soete, 1988; Dzidonu, 2001; McCormick and Onjala, 2007). The ICT policy and regulatory process are being developed and strengthened, and there is evidence of emergent ICT production and services and their utilization and application in various sectors of the African economies. As observed by Wangwe (2007) and Gillwald and Adam (2007), ICT policy and regulatory process have undergone considerable development in Africa in the last two decades, and telecommunication policies and national ICT policies have been implemented as the main instruments of ICT development.

Nigeria has developed a 'National Policy for Information Technology' (ICT policy) and a 'National Telecommunications Policy' (Telecom Policy), both of which are currently being implemented. The telecom policy was launched in 1999 and a National Communication Commission (NCC) was subsequently established to implement the policy. The main objective of the telecom policy is to achieve the modernization and rapid expansion of the telecommunications network and services. This is expected to improve socio-economic development through internal integration of economic and social agents and the integration of Nigeria into the global telecommunications environment. Though the National ICT policy came into effect only in 2001 and is independent of the telecom policy, the technological imperatives of ICTs and telecommunications intertwine the two policies and make their implementation mutually dependent. As reported by Adeyinka and Ubaru (2004), and Adeoti and Adeoti (2008), the implementation of the telecom policy has had considerable impact on access to telephone (especially through licensed mobile telephone operators inclusive of GSM[1]) and the diffusion of Internet technology in recent years.

The implementation of the ICT policy was set in motion in earnest with the establishment of the National Information Technology Development Agency (NITDA) in 2001. The main objectives of the ICT policy are to ensure that ICT resources are readily available to promote efficient national development; and to guarantee that the country benefits maximally, and contributes meaningfully by providing ICT solutions that address the challenges of the information age. Through the implementation of the ICT policy, the usage of ICTs is expected to be pervasive in every sector of the economy with the outcome of increased productivity, economic growth and structural transformation. Besides the rapid increase in teledensity resulting from the liberalization of the telecommunication industry, there has also been significant expansion of ICT services and an emergence of clusters of small and medium sized firms engaged in maintenance and assembly of computers, computer accessories and peripherals. As reported by Oyeyinka (2006) one of these clusters (Otigba ICT cluster) has become well known for the assembly of clones and branded computers which are used in the West African sub-region because they are relatively cheap and well adapted to the local environment. For economic growth and structural transformation, it is imperative that the ICT policy should be a catalyst for the use of ICTs in enterprise management and production systems. Using a survey and case study approach, the study reported in this chapter examines the impact of the ICT policy on firms in the Nigerian printing and publishing industry. The main research question is: 'How effective is the Nigerian ICT Policy in influencing decision to adopt ICTs by the printing and publishing firms?'

2. The Printing and Publishing Industry and ICT Development in Nigeria2

The Nigerian printing and publishing industry

The printing and publishing industry is one of Nigeria's oldest industries. According to Afolabi (2008), the advent of the printing and publishing industry in Nigeria can be traced to the Christian Missionaries Hope Waddel and Samuel Edgerly who established the Hope Waddel and Samuel Edgerly Press in 1846 in Calabar for the purpose of mass production of religious tracts and booklets. This was followed by a similar effort by Rev. Henry Townsend who established a Missionary Press at Abeokuta in 1854. Subsequently, as education and enlightenment advanced in the colonial era, The Daily Times of Nigeria Plc was incorporated in 1925 as the Nigeria Printing and Publishing Company Limited by private individuals. The company was established to carry on business as printers and publishers, advertising agents and newsvendor of a Newspaper to be called The Nigerian Daily Times, and any other newspapers, periodicals and journals.[3]

The printing and publishing industry in Nigeria produces a variety of outputs, including magazines, books, newspapers, and directories. It also produces greeting cards, databases, calendars, and other published materials. Although mostly producing printed materials, the publishing industry is increasingly producing its material in other formats, such as audio or video discs (e.g. CD-ROM), or other electronic media. Educational, political and religious literature, as well as newspapers constituted the bulk of the output from the printing and publishing industry in Nigeria for nearly a century. The industry relies mostly on the enormous needs for school books and teaching materials. With the huge demand[4] for printed books the industry has had to improve on its methods, equipment and skills (Osinowo, 2008; Iyiola, 2008; Orija, 2008).

The printing and publishing industry in Nigeria is a sector to watch out for in terms of the use of ICTs, and the impact this use may have on the productivity of the industry. For example, Ehikamenor (2002) and Bakare (2006) have shown how ICTs opened up new opportunities for the Nigerian print media to improve on their products and services in their studies on the socio-economic factors associated with the adoption and use of ICTs by the print media. According to the findings of the study, the factors that have positive influence on the adoption of ICTs included organizational goal, profitability, organizational image, communication in the organization, productivity, and openness of workers to change.

The Nigerian printing and publishing industry is characterized by the micro, small and medium scale entrepreneurs on one hand and the large scale, mostly foreign owned publishing firms on the other. While many of the former still rely on the traditional printing methods like typesetting and lithography, the larger firms thrive on the use of cutting edge technologies in their production and marketing activities. An important characteristic of the printing industry in Nigeria is the spatial clustering of these outfits. In most cities in Nigeria, a particular area is known for an unusually high concentration of printing presses and related business such as binding, paper cutting, etc. These areas usually attract pool of labour with technical skills in printing and image production. In Lagos, Mushin is noted for the localization of small and medium-sized printing firms. In Ibadan, many of the large publishing firms are located in Jericho area while the small and medium-sized printing presses are mostly located in Mokola area.

The development of ICTs in Nigeria

The development of the ICT sector in Nigeria has proceeded with a focus on liberalization, deregulation and competition in service delivery. ICT policy formulation and implementation in Nigeria is an assignment undertaken in each of the domains that make up the sector. These are the broadcasting, telecommunications and information technology domains. The Wireless Telegraphy Act of 1961 was the main regulatory instrument of broadcasting activities in Nigeria until 1992. Decree No.38 of 1992 and, subsequently its Amendment Decree No. 55 of 1999, established the National Broadcasting Code to deregulate the Nigerian broadcasting industry, and to bring an end to over 50 years of government sole ownership of broadcasting (Amana, 2000).

The Nigerian Communications Commission (NCC) was established by Decree 75 of 1992 to remove the monopoly of NITEL[5] and to facilitate the entry of the private sector into telecommunications service delivery in Nigeria. The creation of NCC was to separate regulatory and operational functions in the sector in response to some of the conditions, amongst others, set out in the ITU African Green Paper for Telecommunication Policies in Africa to facilitate Africa's telecommunication sector deregulation and enhance competition as well as future planning in selected countries. The ITU African Green Paper was formally launched at the second African Regional Telecommunication Development Conference in Abidjan in 1996 (ITU, 1996). The paper set out to explore specific themes related to the overall restructuring of the telecommunications sector in Africa.

NCC was charged with the implementation of the national telecommunications policy that was formulated and approved in 1998 to further liberalize the telecommunications sector. The liberalization thrust of the policy was strengthened by the NCC Amendment Decree No. 30 of 1988, which deleted those provisions in the former decree that inhibited competition in the sector thus enhancing the anticipated role of private sector enterprises. A communications law was signed as the NCC Act in July 2003, repealing Decree 75 of 1992 and its amendment Decree 30 of 1998. The Act is consistent with worldwide standards and grants more powers to the NCC to perform its regulatory role. The new Act gives absolute independence to the NCC in making telecommunications industry regulations. The Act harmonizes the NCC Decrees 75/98 and the Wireless Telegraphy Act.

Realizing the importance of ICT in socio-economic development, the government of Nigeria set in motion strategies that would ensure the effective participation of the country in the new information economy. To achieve this objective, government established the National Information Technology Development Agency (NITDA) in 2001 to implement all policies related to the development and use of ICTs in the country.[6] The National Policy on Information and Communication Technology had earlier in 2001 been formulated and approved to address the issues of low level of ICT development in the country as well as the issues of access and utilization which are inextricably linked to service availability and affordability. In Nigeria, existence of Telecommunications Policy alongside the ICT Policy is particularly noteworthy.

Rowland (1996) (cited in Hafkin, 2002) generally categorized ICT policies into vertical, infrastructural, and horizontal policies. According to him, vertical ICT policy addresses sectoral needs, such as manufacturing, transportation, agriculture, commerce and tourism. Infrastructural aspect deals with the development of national infrastructure that facilitates communications and linkages among the actors in a National Innovation System (NSI). The horizontal aspect deals with the impact on broader aspects of society such as freedom of information, tariff and pricing, privacy and security. The Nigerian IT policy is a comprehensive document that embraces all the key components of the vertical, infrastructural and horizontal policies.

The Nigerian National Policy for Information Technology (IT Policy) is envisioned to make Nigeria an IT-capable country in Africa and a key player inthe Information Society by the year 2005, using IT as the engine for sustainable development and global competitiveness (FGN, 2001). The mission is to facilitate and encourage the use of information technologies in all facets of national life including, education, creation of wealth, poverty eradication, job creation and enhancement of global competitiveness. Apparently, the overall policy target for 2005 has been missed and a new policy agenda is currently being set under Nigeria Vision 20:2020.[7] The Nigerian IT policy benefitted immensely from the ideas generated in the master plan for the development of a national ICT program "ICT 2000" and the first African Development Forum on the Challenge to Africa of Globalization in the Information Age held in Addis Abba in October 1999. Part of the 31 general objectives of the ICT policy include establish appropriate institutional framework for policy implementation; create IT awareness and ensure universal access in order to promote IT diffusion in all sectors; ensure that Information Technology resources are readily available to promote efficient national development; and to build a pool of IT literate manpower. The policy is comprehensive, bringing all facets of the national life under coverage. The ICT for Development (ICT4D) Plan has also been developed as the National Strategic Action Plan for the implementation of the Nigerian IT Policy. The ICT4D plan provides concrete strategies for the application and use of ICTs in key sectors of the economy.

By 1994, there were more than 200 registered companies in Nigeria offering a broad range of ICT related services. A partial survey conducted by Jidaw Systems[8] in 1986, showed that 88 per cent of these were vendors; 79 per cent were consultants; 70 per cent offered training services; 68 per cent had maintenance facilities; while 45 per cent offered bureau services. Presently, the structure of the industry would still be largely the same. Vendors representing all the major global brands in ICT dominate the industry. Between 1999 and now, some of the leading brands have opened country offices alongside their vendors/distributors. Microsoft, Hewlett Packard, Samsung, Dell, CISCO, ORACLE, are among the leading brands with a Nigerian office. The ICT supply scenario until about the end of the 1990s was such that Nigeria imported 100 per cent of all its IT equipment, and a diversity of firms exists to supply, service, and maintain imported equipment. Generally speaking, the Nigerian electronics manufacturing industry which thrived in the 1970s has virtually collapsed. However, starting from about 2000/2001, two wholly Nigerian owned firms, namely Zinox Computers and OMATEK Computers, began to assemble their own brands of Desktop and Laptop computers mainly for the local market and with government patronage.

Omatek Computers was established in 2000 by its mother company, Omatek Ventures Ltd, which has been in existence since 1986 as a computer training outfit, marketer and distributor of branded computer and accessories. By 1990, the company had become a major reseller for Compaq, IBM, ACER and Apple, among other world recognized computer brands. As a result of high sales volumes achieved, the company was appointed a premium partner for Compaq, senior partner for IBM, Apple and Microsoft for turnovers in excess of $1 million annually. In 1993, Omatek Ventures set up its first factory for the local assembly of computers from completely knocked down parts. Omatek Computers is a joint venture project of Omatek Ventures Ltd, Zenith Bank and Guaranty Trust Bank and according to its management, the company was committed to making computers cheaper and more affordable in Africa and also assisting governments to bridge the digital divide, which poses a major policy challenge. The company's products include fully-built desktop PCs, notebooks, casings, and speakers. It is also involved in the assembly of OEM brands for both local and foreign builders. The company uses technology from China, Singapore and India to implement OEM initiatives that are promoted by Intel and Microsoft. The company has a strong international quality consciousness for Omatek brand of products, and has succeeded in receiving more than 18 awards that signify the wide acceptance of its products. The company has also become the first computer firm to be listed on the Nigerian Stock Exchange, and has grown into a Group of Companies with the establishment of subsidiaries to strengthen its operations. Omatek Ventures Plc has therefore become the holding company with subsidiaries that include:

- Omatek Computers Ltd
- Omatek Computers (Ghana)Ltd
- Omatek Ventures (Ghana) Ltd
- Omatek Engineering Services Ltd
- Omatek Ventures Distribution Ltd

Zinox Technologies Ltd was launched in 2001 by a team of IT professionals for the production of computers locally. Zinox claimed to engage in significant research and development as part of its core operations from where products designed for the local markets are conceived, and innovative and customer-centric packages and products are designed particularly for the African market. Zenox was the first internationally certified branded computers OEM in West Africa. It is also the first computer company in the world to incorporate the Naira on its keyboard; and it aims at developing other digital solutions in areas other than computer manufacturing. Zinox has been reported to commence strategic investments in leading ICT companies and has made major investments in Technology Distributions Ltd, Task systems Ltd, AfriHub LLC, and Zinox Telecommunications Ltd. Both Omatek and Zenox brands have enjoyed favourable patronage from public and private sector agencies in Nigeria. The Federal Government in particular has an official policy that encourages its agencies to use Omatek and Zenox brands.

3. Data Sources

The data employed in this study is obtained from a survey of firms in the printing and publishing industry in Nigeria, and three case studies that provide detailed illustration of the impact of ICT policy that may not be adequately captured by quantitative analysis. The data generated from the survey has been elaborately described by Adeoti et al (2010). A small section of the data is used in this study to analyze the impact of ICT policy on a sample of firms in the Nigerian printing and publishing industry. The data is a cross-sectional data obtained by the authors in a survey carried out between the months of March and May 2009.

4. Results and Discussion

Firm size distribution of the research sample

Table 1 shows the distribution of the research sample by firm size. Only three of the firms are large sized firms, two of which are very large publishing firms each employing over 200 persons. Most of the firms employ less than 10 persons, 18% of the firms employ 10-19 persons, while 15.3% employ 20-49 persons. In effect, about four-fifth (80.6%) of the firms employ less than 20 persons, indicating that the research sample comprises largely small-sized enterprises.

Table 1: Distribution of the research sample firms according to size

No. of firms employing	Frequency	Percentage
5-9 persons	40	55.6
10-19 persons	18	25.0
20-29 persons	8	11.1
30-39 persons	1	1.4
40-49 persons	2	2.8
50 or more persons	3	4.2
Total	**72**	**100.0**

Source: Adeoti et al (2010, p.258)

Nigerian ICT policy impacts

Awareness of ICT policy

To a large extent, the assessment of the effectiveness of the ICT policy is expected to start from the awareness of the existence of the policy. Respondents in our survey of printing and publishing firms were therefore requested to indicate whether or not they were aware of the existence of the Nigerian ICT policy. The results revealed that about 68 per cent of the respondents reported that they were aware of the ICT policy, while the remaining 32 per cent said that they were not aware. Awareness of the Nigerian ICT policy in the printing and publishing industry, as typified by our sample respondents can be conjectured to be substantial, and this paints a good picture for the prospects of ICT use in this industry. The expectation is that awareness of the existence of the ICT policy will help firms to take advantage of the challenges and opportunities that come with the adoption of ICTs.

Impact of ICT policy on firm performance

The respondents who claimed to be aware of the ICT policy reported on the varied impact which the policy has made on their firm's performance especially in the last three years. The responses in this respect are reflected in Figure 1. Taking a general outlook, it would seem that the ICT policy had impacted at least 40 per cent of the respondents of this survey more in the positive manner, such as improvement in market for products, introduction of new or improved products, adoption of new equipment/machines, made investment in ICTs less less expensive, and stimulation of more investment in ICTs. Specificallly, 73.1 per cent of the respondents claimed that the market for their products had improved, while 61 per cent said it had enabled the introduction of new or improved products. Close to 50 per cent claimed to have adopted new equipment/machines, about 44 per cent said it made investment in ICTs less expensive, while 40 percent claimed that it stimulated investment in ICTs.

Figure 1: Impact of ICT policy on firm performance

Source: Analysis of survey data

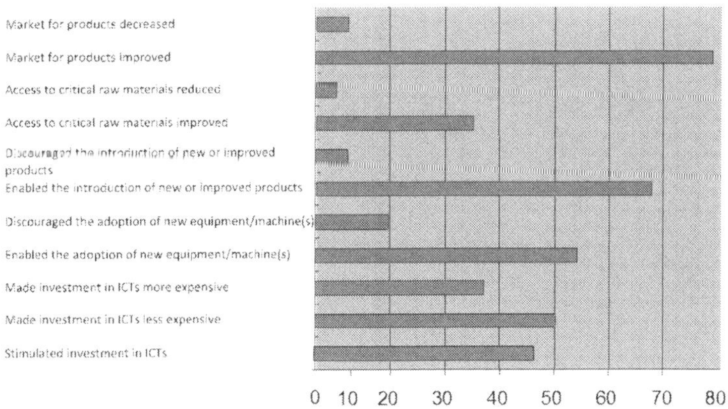

Respondents were asked to identify the two most important impact that the

ICT policy had made on their firm in the last three years. The responses on this question are shown on Table 2. Consequently, an average value was calculated to give a rating to the impacts in an ordered manner. The two most important impact of ICT policy on the performance of the research sample firms in the last three years are thus identified as "it has enabled the introduction of new or improved products" and "it has enabled improved market for products".

The findings reported in the foregoing conform to the widely-held belief that ICTs are important enablers of growth through improved business performance. For example, previous studies such as Bresnahan et al. (2000) andMouelhi (2008) submit that adoption of ICTs (Internet, mobile telephony and broadband networks) in many developed countries has been found to have positive effect on organizational performance and firm growth. It was reported that, by tackling the barriers of time and distance, ICTs contribute immensely by providing efficiency gains and increased productivity.

Table 2: Ranking of the impact of ICT policy on firm performance

Type of impact	% Respondents that claim impact		
	Most important	2nd most important	Average
Stimulated investment in ICTs	5.9	6.1	6.0
Made investment in ICTs less expensive	17.6	10.2	13.8
Made investment in ICTs more expensive	9.8	4.1	6.95
Enabled the adoption of new equipment/ machine(s)	9.8	12.2	11.0
Discouraged the adoption of new equipment/machine(s)	0	4.1	2.05
Enabled the introduction of new or improved products	19.6	14.3	16.95
Discouraged the introduction of new or improved products	0	6.1	3.05
Access to critical raw materials improved	5.9	12.2	9.05
Access to critical raw materials reduced	3.9	24.5	14.2
Market for products improved	27.5	6.1	16.8
Total		100.0	100.0

Source: Analysis of survey data

However, it is important to note that there are no previous empirical studies that clearly suggest or signify the effectiveness of the Nigerian policy on ICT. Except for the widely acknowledged success recorded in the Nigerian telecommunication industry (Adeyinka and Ubaru, 2004; Adeoti and Adeoti, 2008), the effectiveness of the Nigerian ICT policy is perceived to be limited in impact and scope. Like in many other sub-Saharan African countries, the penetration and usage of ICTs are still relatively low in Nigeria. Butcher (2003) identified the factors responsible for low level of ICT penetration and usage in Africa to include: weak telecommunications infrastructure to support the rapid development of ICT in the continent; restrictive institutional structures which are ill adapted to facilitate these developments to meet the huge demand for ICT services; inadequate regulatory systems and frameworks to guide the development of the sector and rapid diffusion of ICTs; weak and non-existent regional links to help create economies of scale and drive cost of capital equipment down; general low level of economic activity; irregular or non-existent electricity supplies; lack of related infrastructure (e.g., electric power) needed to implement and support ICT infrastructure; and lack of highly skilled human resources (exacerbated by brain drain) to support roll out, design and exploitation of ICTs. Moreover, there is poor local capital market to facilitate the mobilization of funds to support ICT developments, poor commitments by leadership within African countries to drive ICT investment, and there remains the perception of Africa as a high risk continent with weak governance structure and systems.

On the supply side, the harsh manufacturing environment in the country has prevented foreign direct investment (FDI) inflow especially in the areas of ICT equipment manufacture. It is doubtful if the ICT policy has been able to make significant impact in addressing this problem. At the time of this study, there are only four firms engaged in local manufacturing of computer hardware while the software industry still remains relatively under-developed when compared with that of emerging economies such as India and China. The computer manufacturing firms include: Zinox Technologies Ltd. Lagos, jointly owned by Stan Tech, Nigeria, Mustek, South Africa, and Alhema, France; United Information Technologies (UNITEC); Omatek Computers; and Beta Computers. As reported by Oyeyinka (2006), these firms have entered the Nigerian manufacturing market with different products. The Zinox Computers have a number of components and parts (e.g., power circuits, casing, keyboard, and packaging) fabricated abroad to the company's design. This has caused a near total dependence on imported components and software. Nigeria still spends billions of Naira annually on imported hardware and software, attesting to unimpressive performance of the ICT policy in encouraging local production of parts. Moreover, the global downward trend in the pricing of ICT products, and particularly the influx of cheap computers and allied products from China and other Newly Industrializing Countries of South East Asia pose great challenges to the effectiveness of the ICT policy in stimulating local manufacture of ICT products. The local firms mentioned above offer ICT products at prices slightly higher than the imported equivalents, understandably due to generally acknowledged relatively high cost of manufacturing in Nigeria.

Case study illustrations

In order to deepen insight into the pattern of use and the specific impact of ICTs in the Nigerian printing and publishing industry, three firm-level case studies were conducted. The basic features of the three selected firms are shown in Table 3, and they may be categorized as:

Firm A: Small-sized firm, late adopter of ICT

Firm B: Small-sized firm, early adopter of ICT

Firm C: Medium-sized firm, early adopter of ICT

The firms were visited by the researchers, who conducted in-depth interviews with the owners/managers of the firms. The findings are presented as follows with a focus on challenges and awareness of ICT policy; nature of outputs and printing technology; and pattern of ICT use and impacts.

Table 3: Basic characteristics of case study firms

Characteristic	Firm A	Firm B	Firm C
No. of persons employed	8	25	31
Turnover in 2008 (million Naira)	4.3	9.7	25.9
Capacity utilization in 2009 (%)	40	60	50
Affiliate of MNE?	No	No	Yes
Age of firm (yrs)	44	11	16
Age of main production equipment (yrs)	16	8	2
Total book value of machines/equipment in 2008 (million Naira)	2.5	20.0	42.0

Source: Field data

Firm A: Small-sized firm, late adopter of ICT

Challenges and awareness of ICT policy

The firm is a relatively very old printing firm established in 1965 and employs eight people when in full operation. The owner is one of the pioneer printers in Ibadan. The press is currently under the management of one of the owner's son who undergone apprenticeship in the firm. Up to the late 1980s the firm had a thriving printing and publishing business. However, the firm has been faced with the challenges of inadequate power supply and proliferation of the printing business beginning from the 1990s when desktop computers became common resulting in many untrained people claiming to be printers. This problem has been compounded in recent years by lack of patronage from big publishers who formerly sub-contracted printing jobs to the firm. It is now the practice of big publishers to send the printing jobs to Asia, especially India, because it is considered cheaper to print abroad. Cost of generating electric power by diesel generator has made production too expensive. Power supply from PHCN[9] is at an average of only two hours out of eight production hours per day in 2008. The capacity utilization at the press has continued to decline in recent years: from 90% in 2006 to 70% in 2007, 50% in 2008, and 40% in 2009. The firm has been slow to invest in computer, and the owner still lacks computer skills. The firm is unaware that ICT policy exists and thus is uninformed about its contents. The firm believes that the government should address the basic problem of electric power supply; regulate the activities of printers that lack professional training and printing skills; and ban the offshore printing by publishers.

Nature of outputs and printing technology

The firm is a general printing firm and its outputs include bulletins, exercise books, posters, handbills, stickers, calendars, invitation cards, and books. Its annual sales turnover was N4.3million in 2008. The firm is involved in all the three stages of printing: pre-press, press, and post-press. The firm has one half-cord multigraphics offset printing machine and a big-size cutting machine. The offset machine was purchased two years ago but it is actually 16 years old, imported as a refurbished second hand machine from Germany. The printing technology used includes lithography and digital printing. The offset machine is capable of printing only up to A3 sizes while the cutting machine is one of the biggest available in Ibadan. The firm contracts jobs involving sizes bigger than A3 to bigger presses while it also engages in cutting papers for other presses.

Pattern of ICT use and impacts

The firm owns a desktop computer, which was purchased out of the concern that new generation printers who are skilled in using computers have made competition difficult and challenging. None of the firm's workers received additional training in 2008 and the firm has never trained any of its workers in ICT skills. When the need arises, a computer literate person is employed on ad hoc basis to operate the firm's desktop computer. The most important factor that has discouraged investment in and use of ICTs by the firm is the high cost of ICTs, followed by lack of information on relevant ICTs. The firm does not use any e-business facility and by implication lack IT skills required for using the Internet or online interaction with clients. The owner of the firm is the patron of master printers association in Ibadan, and complained about lack of unity among printers basically arising from rivalry and impact of ICTs that has made people unskilled in printing to emerge as printers. It is contended that desktop publishing skills do not turn a person to professional printer. A printer must be skilled in the actual printing operation especially at the press and post-press stages.

Firm B: Small-sized firm, early adopter of ICT

Challenges and awareness of ICT policy

This firm was established by a young entrepreneur in 1998. The owner obtained a university degree in Agricultural Economics in 1990 but chose to pursue a career in printing while still in university.[10] The firm employs 25 persons and runs two production shifts. The firm has been growing especially in recent years and capacity utilization has improved steadily since 2006: 30% in 2006, 38% in 2007, 45% in 2008, and 60% in 2009. The firm is aware that the ICT policy exists but is uninformed about its contents. The most crucial challenge facing the firm is lack of power supply from PHCN, which provides power at an average of only 25 minutes out of eight production hours in 2008. Another important challenge for the firm is difficulty in accessing latest printing technologies and state-of-the-art equipment. According to the firm, these are very expensive because they often require overseas training and high capital outlay. Those that have the training are reluctant to share the knowledge for competitive reasons. However, the firm strives to acquire knowledge as much as possible especially in the area of application of ICTs. For instance, when digital separation came to Nigeria about seven years ago, the entrepreneur spent three days in Lagos to learn how to employ the new technology. He also occasionally embarks on short courses abroad to update his knowledge on printing technology. He claims that it is discouraging to professional printer that government policy allows some big publishers take their manuscripts abroad for printing while there are local firms that can produce equally good or even better products. He suggests that the ICT policy with respect to printing should aim at empowering the local firms by banning offshore printing of local manuscripts. He also claimed that the volatility in exchange rate also poses a challenge for the firm's sourcing of largely imported raw materials, and this creates uncertainties and eventual high cost of production. He stated that most of the machine spare parts, papers and ink used in the Nigerian printing industry are imported especially from Germany and China.[11] A major challenge of ICT policy should therefore be how to ensure that good spare parts and high quality raw materials for printing are readily available and produced locally.

Nature of outputs and printing technology

The firm is a general printing establishment and its outputs include books, calendars, posters, handbills, stickers, invitation cards, ceremonial programmes, etc. The firm boasts of products that are of international quality standards. It has printed books for clients in Ireland, New Jersey in the USA, and Manchester in the United Kingdom. The firm's sales turnover was N9.7million in 2008. The firm engages in all stages of printing: pre-press, press, and post-press. The firm is involved in lithography, screen printing, and digital printing. The firm has three heavy duty full-cord offset printing machines. The oldest was purchased eight years ago, while the latest was bought three years ago.

Pattern of ICT use and impacts

The firm believes that it is important to be proactive in taking advantage of opportunities presented by ICTs in printing technology. In the last three years the firm has invested in purchase of new computers and software for desktop publishing, and electronic inventory monitoring system to manage the increasing volume of work. The firm has employed one ICT expert to work full-time on ICT related jobs, and the firm has successfully employed ICTs in the production process, office automation, customer relations management and management of product distribution network. The firm uses e-mail often to communicate with clients and occasionally engages in online website enabled transaction. The internet has been used by the firm as source of information on new printing technologies and as source of very useful graphics/pictures. The factors that have generally favoured investment in ICTs by the firm include production process improvement, competition among printing and publishing firms, and product quality improvement. The main factor that has limited the firm's investment in ICTs is the high cost of such investments.

The use of ICTs by the firm has had advantages and disadvantages. On one hand, it has provided tremendous speed for job production and delivery, while on the other hand, it has made the firm's products more vulnerable to piracy. As an example to illustrate ICT advantage, the firm stated that a picture to be used in a poster can be sent via the internet and downloaded in few minutes. Necessary desktop work can then easily be done and print output obtained with incredible speed. This would have saved time and substantial costs that could have been involved without ICT. The firm thus claimed that, to improve efficiency and remain competitive, it uses ICTs at every stage including packaging, processing, litho graphics, printing, pre-press, post-press and scanning. The firm believes that the use and impact of ICTs in the Nigerian printing industry would be further enhanced if a university degree programme in printing technology is introduced in Nigeria.[12]

Firm C: Medium-sized firm, early adopter of ICT

Challenges and awareness of ICT policy

The firm was established in 1993 and is affiliated to Eduserve Consult, which is an educational consultancy and outreach outfit.[13] In the context of this study, the firm is a medium-sized printing press employing 31 persons. The firm's capacity utilization has ranged between 40 and 50% in the last three years. The main challenges confronting the firm are the grossly inadequate power supply from PHCN; and relatively high tariffs on imported printing machines and raw materials such as paper and ink. The firm is usually on own electricity generator. For example, it was reported that power supply from PHCN is at an average of only one hour out of ten production hours per day in 2008. The firm is aware of the ICT policy, and claimed that the policy has encouraged its investment in ICTs, resulting in improvement of market for its products.

Nature of outputs and printing technology

The firm boasts of being one of the foremost Book Publishers and Press in Nigeria, and its sales turnover was almost N26million in 2008. The firm is involved in all stages of printing: pre-press, press and post-press. The firm has modern and relatively new offset printing machines, the latest was acquired two years ago from Germany. The printing technology used by the firm includes lithography and digital printing. The owner/manager of the firm is an editor who had worked in an international printing and publishing company where she acquired skills in printing technology. The firm recently bought a new ruling machine in 2008, uses the latest state-of-the-art scanners to improve print images, and strives to maintain high production standards. To reduce costs of paper as a major raw material, the firm tries to be innovative by cutting large-sized papers into smaller sizes instead of buying small sized papers. For this purpose, the firm had to invest in a cutting machine. The press does not have machines to print sizes larger than A2. Jobs involving A1 size are subcontracted to other firms. The firm gives priority to employing people who are professional printers with evidence of completion of a course in printing technology or apprenticeship. The firm frequently receives contract from other publishing companies that concentrate on a particular niche of publishing. For example, a large publishing company like Evans that specializes in the pre-press stage often engages the services of the firm to undertake the press and post-press stages.

Pattern of ICT use and impacts

The main types of ICT that the firm has invested in, especially in the last three years are computers and computer software packages, such as Corel Draw. ICT is employed by the firm in office automation, production processes, customer relation management and management of distribution networks. The main type of e-business facility regularly used by the firm is the electronic messaging system (Email). The firm does not engage in online portal-based transaction but occasionally engage in online website enabled transaction with clients. The firm has in its employment two persons with ICT skills devoted to working on desktop publishing and other ICT related aspects of the press. In 2008, ICT consultants were engaged to further train the ICT staff, editors and lithographers in-house. For example, they were trained on the use of the latest updates in frequently used software such as Corel Draw. The factors that have favoured the firm's investment in ICT are the opportunities for improving the product quality through ICT applications and the challenge of competition among the printing and publishing firms. The main factor that has discouraged or limited the firm's investment in ICTs is the high cost of implementing ICT solutions.

The firm claimed to be a recognized and valuable member of international and national industry associations such as the Nigerian Publishers Association, ABC (UK), and Association of West African Book Editors. The firm claimed that industry associations have not really taken up the challenge of assessing the implementation or impact of the ICT policy, but this is "food for thought", and it would be a good theme to focus on in future. Nonetheless, in the past, these associations have discussed on specific topics such as impact of ICTs on copyright issues, education and the way forward, piracy, and trends in printing and publishing technology.

5. Concluding Remarks

From the findings of this study it is evident that most of the firms in the research sample have adopted ICTs and are employing ICTs in various activities and operations. 68% of the respondents are aware of the existence of the ICT policy, but some of them are ignorant of its contents. This is an indication that there is need for more awareness campaign to enable stakeholders in the Nigerian printing and publishing industry gain insight into the ICT policy and take advantage of the opportunities promoted by the policy.

The respondents who claimed to be aware of the ICT policy reported varied impacts of the policy on firm performance. The results showed that the ICT policy had impacted at least 40% of the respondents in a positive manner, such as improvement in market for products, introduction of new or improved products, adoption of new equipment/machines, made investment in ICTs less expensive, and stimulation of more investment in ICTs. This not withstanding, the case study illustrations suggest that the ICT policy impact is still not deep, and firm expectations in the printing and publishing industry are still far from being met. It would therefore be necessary to further strengthen the implemetation of the ICT policy as it applies to the priniting and publishing industry.

Notes

1 Global System of Mobile Communication (GSM).

2 This section draws largely from Adeoti et al. (2010).

3 The company started printing on 1st June, 1926 with the publication title, 'The Nigerian Daily Times'. See http://www.dailytimes-nigeria.com/ (Accessed: 25 February 2009).

4 Over 95 per cent of books used in the primary and junior secondary schools are locally printed (written, edited and illustrated by the Nigerians). Nigerian publishers are now going into the senior secondary school sector and into the technical, professional and tertiary sectors of textbook production. The estimates made in the mid-1980s showed that a minimum of 285million textbooks per annum at all levels of education would be needed (Culturelink, 1996).

5 NITEL was the government/state telecommunications monopoly until 1992 when the sector became liberalized.

6 The Act for the establishment of NITDA was signed into law in March 2007.

7 Vision 20:2020 is Nigeria's long term development agenda aimed at making Nigeria one of the 20 largest economies by the year 2020.

8 See http://www.jidaw.com/itsolutions/policynigeria.html (Accessed: 24 February 2009).

9 PHCN is Power Holding Company of Nigeria, the sole electric power utility company.

10 As a university student he undergone apprenticeship in printing on a part-time basis and was able to support himself through income earned as a student-printer. After his graduation he worked as free-lance printer until he was able to save up to N335,000.00 which served as his start-up capital in 1998.

11 The entrepreneur emphasized that the Chinese copied the printing technology from Germany and now they are expert in the production of printing papers; and most of the raw materials used for printing in Nigeria are imported from China.

12 In Nigeria, the highest formal educational training in printing technology has been a Higher National Diploma in printing technology at the Yaba College of Technology, Lagos and Kaduna Polytechnic, Kaduna. A private university in Ibadan however recently started a degree programme in printing technology.

13 Eduserve Consult is the proprietor and founder of this printing and publishing firm and the firm is a sister organization to a private university also established in Ibadan by Eduserve Consult.

References

Adeoti, J.O. and A.I. Adeoti (2008), 'Easing the Burden of Fixed Telephone Lines on Small Scale Entrepreneurs in Nigeria: GSM Lines to the Rescue', in *Telematics and Informatics*, 25(1):1-18.

Adeoti, J. O., Adeyinka, F. M. and Ubaru, M.O. (2010), 'Efficiency Impacts of Information and Communication Technologies in the Nigerian Printing and Publishing Industry', in *Int. J. Technological Learning, Innovation and Development*, 3(3): 245–271.

Adeyinka, F. M. and Ubaru, M. O. (2004), 'Strengthening Information and Communication Technology Policies in Africa: Governance and Equity Issues - The Case of Nigeria', *ATPS Research Report*, Nairobi, Kenya: African Technology Policy Studies (ATPS) Network.

Afolabi, A. (2008), 'ICT: Driving the print media industry in Nigeria', in *Technology Times*, Tuesday, 3 June 2008. Available at: www.technologytimes.com (Accessed: 10 June 2009).

<small>Amana, E. I. (2000), 'Broadcasting in Nigeria', in *Proceedings of Workshop on National Information and Communications Infrastructures (NICI) Policy, Plans and Strategies*, Held at National Universities Commission Auditorium, Abuja, March 28-30.

Bakare, M. (2006), 'Publishing Literature in Nigeria - Challenges and Opportunities', Paper presented at the African Studies Association of the UK (ASAUK) Biennial Conference, *School of Oriental and African Studies (SOAS), London, 11th-13th September.*

Bresnahan, T.F., Brynjolfsson, E. and Hitt, L. M. (2000), 'Information Technology, Workplace Organization and the Demand for Skilled Labor: Firm-level Evidence', *Discussion Paper*, Department of Economics, Stanford University, Stanford, CA.

Butcher, N. (2003), 'Technological Infrastructure and Use of ICT in Education in Africa: an overview', *Report of the Working Group on Distance Education and open Learning (WGDEOL)*, Association for the Development of Education in Africa, Tunis, Tunisia: African Development Bank (AfDB).

Culturelink (1996), 'Cultural Policy in Nigeria', in *Compton's Interactive Encyclopaedia*, Carlsbad, CA: Compton's New Media, Inc.

Dzidonu, C. (2001), 'A framework for guiding the development of ICT-led socio-economic development policies, strategies and plans', *Report commissioned by the Africa Technology Policy Study*, Accra, Ghana: International Institute for Information Technology (INIIT).

Ehikhamenor, F.A. (2002), 'Socio-economic factors in the application of information and communication technologies in Nigerian print media', in *Journal of the American Society for Information Science and Technology*, 53(7): 602–611.

FGN (2001), *Nigerian National Policy for Information Technology*, Abuja, Nigeria: Federal Ministry of Science and Technology.

Gillwald, A. and Adam, L. (2007), 'The Political Economy of ICT Policy Making in Africa: Historical Contexts of Regulatory Frameworks, Policy Performance, Research Questions and Methodological Issues', *Paper prepared for a special research project by African Economic Research Consortium, (Nairobi) on ICT Policy and Economic Development in Africa*, Nairobi, Kenya: AERC.

Hafkin, N. (2002), *Gender issues in ICT policy in developing countries: An overview*, United Nation's Division for the Advancement of Women (DAW) Expert Group meeting on Information and Communication Technology and their impact on and use as an instrument for the Advancement of Women, Seoul, Republic of Korea. http://www.apc.org/english/capacity/policy/#mmtk_gender_ictpol_hafkin.pdf (Accessed: 1 June 2009).

ITU (1996), *The African Green Paper: Telecommunication Policies for Africa 1996*, Geneva, Switzerland: International Telecommunication Union (ITU).

Iyiola, J. (2008), 'Continued education and constant training boost the growth of a company', *Paper submitted to Print Promotion Writing Contest on the occasion of DRUPA 2008*, under the theme "Print Promotion, My Work, My Passion, My Life", 29 May – 11 June, Dusseldorf, Germany.

McCormick, D. and Onjala, J. (2007), 'Methodology for Value Chain Analysis in ICT Industry: Frameworks for the Study of Africa', *Paper prepared for a special research project by African Economic Research Consortium, (Nairobi) on ICT Policy and Economic Development in Africa*, Nairobi, Kenya: AERC.

Mouelhi, R. (2008), 'Impact of the adoption of Information and Communication Technologies on firm efficiency in the Tunisian manufacturing sector: Equity and Economic Development', *ERF 15 Annual Conference*, 23-25 November, Cairo, Egypt.

Orija, O. (2008), 'Training, training, training is the mantra often repeated by press manufacturers and printers too', *Paper submitted to Print Promotion Writing Contest on the occasion of DRUPA 2008*, under the theme "Print Promotion, My Work, My Passion, My Life", 29 May – 11 June, Dusseldorf, Germany.

Osinowo, O.A. (2008), 'The future of the printing industry in Nigeria is bright', *Paper submitted to Print Promotion Writing Contest on the occasion of DRUPA 2008*, under the theme "Print Promotion, My Work, My Passion, My Life", 29 May – 11 June, Dusseldorf, Germany.

Oyeyinka, O. O. (2006), 'Learning Hi-Tech and Knowledge in Local Systems: The Otigba Computer Hardware Cluster in Nigeria,' *Discussion Paper*, DP 2006-07, UNU-INTECH, Maastricht, The Netherlands.

Perez, C. and Soete, L. (1988), 'Catching up in Technology: Entry Barriers and Windows of Opportunity,' in Dosi, G., Freeman, C., Nelson, R., Silverberg, G. and Soete, L. (eds.), *Technical Change and Economic Theory*, London: Pinter Publishers.

Wangwe, S. (2007), 'Evolution, Status and Impact of ICT on Economic Development and Transformation in Africa: An Overview', *Paper prepared for a special research project by African Economic Research Consortium, (Nairobi) on ICT Policy and Economic Development in Africa*, Nairobi, Kenya: AERC.

Acknowledgements

The authors gratefully acknowledge the African Economic Research Consortium (AERC), Nairobi, Kenya for the funding of the research project from which this chapter is drawn. The views expressed in the chapter are those of the authors, and neither the AERC nor any of the authors' institutional affiliation (NISER and NITDA) bears responsibility for them.

Chapter 10
An Assessment of Application Software Market in Cameroon

Sunday A. Khan & Arsène Gidéon Nkama

1. Introduction

It is widely recognized in economic literature that Information and Telecommunication Technology (ICT) plays an important role at both the macroeconomic and microeconomic levels. This is attributed to its rapidly changing nature and its ability to rapidly integrate world economies. The changing nature of ICT keeps an innovative pressure on both intra-ICT industries and in the rest of economy, and therefore facilitates rapid increase in total factor productivity. By increasing globalisation of economies, technology raises the international competitive pressures and then necessitates continuous improvements in the competitiveness of firms. At national level, it is the competitiveness of microeconomic entities that explains macroeconomic growth. Technology is viewed as a major determinant of production efficiency and it is the competitiveness of firms that finally leads to the competitiveness of the nation.

As underlined by Pohjola (2001), ICTs play a dual role in any modern economy: an output from ICT-producing industries and an input to ICT-using industries. As output, ICT production is characterized by fast improvement in the quality of ICT equipment and software and the resulting significant decline in their prices. This fast improvement in quality, combined with the quality adjusted prices leads to an increase of their utilization at both individual and firms levels. The direct impact concerns increase in GDP. Another effect to be considered is the increased utilization of ICT investment at individual and firm levels leading to the increase of the share of ICT in both GDP and total capital stock. ICT is also expected to increase labour productivity even though it can also contribute to the reduction of labour intensity in some sectors. But it should be outlined that ICT is a job-generating industry because of its perpetual innovation that not only needs skilled labour to be sustained, but also lead to new production possibilities, new service provision, etc.

Globally speaking, ICT can enhance output and economic growth in three different ways. First, the production of ICT goods and services contributes directly to the total value added generated in an economy. Secondly, the use of ICT capital as an input in the production of goods and services can accelerate economic growth. Finally, ICT can contribute to economic growth through the contribution of ICT industries to rapid technological change.

According to Wangwe (2007), ICT plays a major role in all aspects of national life: in politics, in economic, as well as in social and cultural development. It transforms people's lives, the way they do business, access information and a service, communicate with each other, provide entertainment and has benefits relating to human rights by supporting freedom of expression and right to information. The Cameroonian society is no exception to these ICT-induced transformations.

It is important to note the apprehension about the impact of ICT on the economy. In the early stages of ICT adoption, as noted by Oyeyinka (2007), there was considerable fear that the adoption of these technologies might result in reduced levels of employment, particularly of semi- or un-skilled workers, due to the programmability feature of ICTs and their capability in handling multiple tasks with a single ICT tool. It is still believed that ICT tools replace certain categories of workers and lead to a significant rise in unemployment levels. According to Nour (2002), this is based on the idea that technical change is a creative-destruction process that creates opportunities for development, while also imposing certain restrictions to development. Such reasoning is valid particularly at the enterprise level with respect to existing manufacturing processes, though the expectation is that the impact on employment at the macro level is positive.

ICT is a convergence of a set of technologies and applications which can be disaggregated into three interlocking elements: telecommunication, computer hardware, and software (Bamiro, 2007). Software can be subdivided into system or operating software, program/proprietary software and application software. The first helps run the computer hardware (MS

DOS, Microsoft Windows ...); while the second are software packages developed to accomplish specific tasks like word processing, spreadsheets, database management, etc. The third helps end users accomplish specific computer related tasks; examples include educational software, management software, accounting software, etc. This study focuses on the last category, application software, which are either tailor-made for individual users or standard packages that can be customised to meet the needs users. The other categories require huge human/financial resources for their development, which are not within reach of many African countries.

Application software is therefore a domain in which countries like Cameroon can contribute to ICT and use it to transform their economies. Unfortunately, to date, nobody can say with certainty the extent to which Cameroon is involved in ICT production (both hardware and software) or what the government is or should do to promote production, especially of application software. In its ICT sector development strategy, the National Agency for ICT (ANTIC[1], 2007: 31-32), indicates that Cameroon is entirely dependent on the external world for its ICT equipment and that the few enterprises involved are mostly local representatives of foreign firms. The government ICT strategy document however focuses almost entirely on the promotion of ICT use. It fails to outline a clear strategy to encourage ICT conception, production and commercialisation, and especially application software. It does not even see the need to assess the level of involvement of Cameroonians in ICT production and look for ways to promote such activities. The main questions for this study therefore are: which are the main application software used in Cameroon, who produces/supplies them (how involved are Cameroonians in software development) and what is the level of competition in this market?

The main objective of the study is to examine the application software market in Cameroon with emphasis on software used in the financial sector. More specifically the study strives to: a) identify the main application software used in the financial sector in Cameroon and the origin of such software – whether locally developed or imported; b) articulate the competitive environment in the financial sector software market; c) examine the level of local application software production/development with the use of case studies; and d) examine the policy options for promoting software development in Cameroon and enhancing its contribution to the growth of the country.

2. ICT Penetration in Cameroon

The political commitment of the government to promote ICT has been demonstrated in several ways, including in a speech by the President of the Republic to youths: "One of the ways of moving into the knowledge economy under construction is to master information and communication technology"[2]. The revised version of Cameroon's PRSP presents a diagnosis of the ICT sector and outlines the country's strategy to enhance the contribution of ICT to growth and poverty reduction. The government's ICT goal is to build a system of production and diffusion of knowledge capable of providing each sector of the economy, at any time and everywhere, the information necessary to transform opportunities in its environment into viable goods and services (Cameroun, no date: 23). A Special Telecommunications Fund has been set up to boost ICT funding. There is therefore no doubt that the government believes in the virtues of ICT and has opted for its vulgarisation in the country.

Access to ICT in Cameroon

According to Nana Nzepa (2003), internet was first introduced in Cameroon in 1991. As shown on Table 1, the number of internet users in Cameroon increased rapidly from less than 4 per 1000 inhabitants in 2003 to over 22 per 1000 in 2007. Compared with other countries, internet use is still very low. In 2003, more than 14 Ivorians and almost 22 Senegalese per 1000 were already using internet, while the same figure in Cameroon was less than 4, much lower than the African average of almost 16 inhabitants per 1000. Four years later in 2007, internet use in Cameroon still remained far below the comparator countries. However, if we compare the rates of increase in the use of internet, we realise that the rate of adoption is very high in Cameroon, given that it started from a lower level. Most Cameroonians access internet through cybercafés which are widespread in the country. There are many internet service providers[3], including the two main mobile phone providers who have made wireless access very easy.

Between 2003 and 2007, the total number of telephone subscribers has rapidly increased as in many African countries. However, the main telephone lines per 100 inhabitants have remained far below the African average in 2007. The country experienced a rapid increase in mobile subscription; however the penetration rate is lower than in Cote d'Ivoire and Senegal, and still has to reach the average African level. There are two private mobile phone operators in Cameroon, Orange and MTN, operating in a duopoly-like market of mobile phones. In fact apart from CT phone of CAMTEL in a few localities, Orange and MTN could be considered as the sole mobile phone service providers.

There are two private mobile phone operators in Cameroon both having a GSM 900 licence: Orange and MTN. Orange Cameroon is a subsidiary of a French company (France Télécom) and started operations in Cameroon in July 1999. Meanwhile MTN Cameroon is owned by South Africans and began supplying mobile phone services in Cameroon since February 2000. These two firms operate in a duopoly-like market of mobile phones. In fact apart from CT phone of CAMTEL in some localities, Orange and MTN can be considered as the only mobile phone service providers. This market has experienced a rapid growth at its early stage, moving from 5 thousand subscribers in 2000 to 700 thousand in 2002; 1.167 million in 2003; 3.135 million in 2005 and more than 4 millions in 2007.

Table 1: ICT Access Indicators in Cameroon and Comparable Countries (2003 and 2007)

ICT indicators	Cameroon		Cote d'Ivoire		Senegal		Africa	
	2003	2007	2003	2007	2003	2007	2003	2007
Total telephone subscribers per 100 inhabitants	5.13	27.47	9.13	23.44	7.77	31.51	8.66	30.94
Main telephone lines per 100 inhabitants	0.7	1.02	1.43	1.41	2.21	2.17	3	3.21
Mobile subscribers per 100 inhabitants	6.62	24.45	7.7	36.6	5.56	29.33	6.18	28.44
Internet users per 1000 inhabitants	3.79	22.3	14.43	16.3	21.7	66.2	15.6	54.6
Personal computers per 100 inhabitants	0.57	---	0.93	---	2.17	---	1.44	---

Source: International Telecommunication Union database

Figure 1: Evolution of Telephone Subscriptions

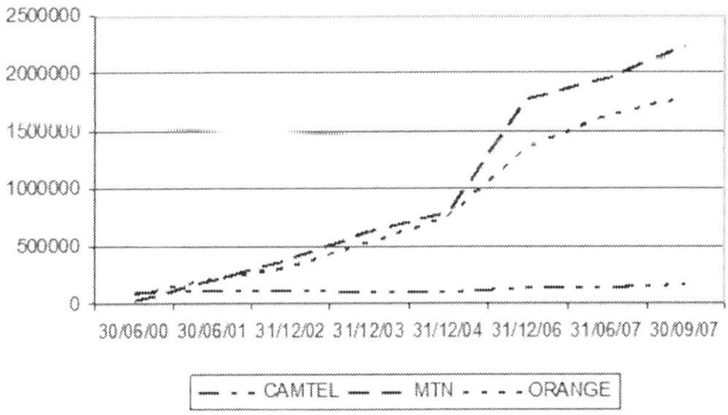

Source: Ngae and Nkolo Essisimi (2000-2004); ART (2006 - 2007)

Affordability of ICT in Cameroon

The relatively low ICT penetration in Cameroon can be explained by many factors of which prices play an important role. According to the World Bank (2008), the price basket for access to internet in 2006 was US$17.6 per month, while the average for Sub-Saharan African countries was US$15.9 per month. The average for all lower-middle income countries was much lower at just US$10 per month.

Table 2: Affordability of ICT Services in Cameroon and Elsewhere in 2006

ICT Services	Cameroon	Sub-Saharan Africa	Lower middle income group
Price basket for fixed line (US$ per month, residential)	9.3	11.6	8.2
Price basket for mobile (US$ per month, 2006)	16.3	12.3	9.8
Price basket for Internet (US$/ month)	17.6	15.9	10.0

Source: World Bank (2008)

We arrive at the same conclusion when we compare the prices of mobile phone services by the same providers in Cameroon and other countries as in Table 3. Orange bills a minute of a local call in Cameroon for CFAF200 but charges CFAF100 and CFAF90 for the same call in Cote d'Ivoire and Senegal respectively. The same is true for MTN services in Cameroon and Cote d'Ivoire, and for international calls.

Table 3: Cost of a cell phone call in Cameroon and some selected countries per Minute

Operators Type of call	MTN Cameroon	Orange Cameroon	MTN Côte d'Ivoire	Orange Coted'Ivoire	Orange Senegal
Intra network call	200	200	100	100	90
Inter network call	200	200	100	150	90
International call	200/300	300/400/500	200/250	300	170

Source: ART (2008)

Efforts are necessary to align phone prices in Cameroon with those in other countries so that phone services can provide an impetus to economic activities in the country. The Telecom Regulatory Board threatened imposing a price cap if mobile operators fail to reduce communication prices (ART, 2008:.20).

Institutional and Regulatory framework of ICT in Cameroon

The government has set up a number of institutions to implement and regulate its ICT policy. The first of these institutions is the Ministry of Posts and Telecommunications which elaborates, implements and evaluates government policy concerning postal and telecom activities. A myriad of other institutions have been set up to accomplish specific missions, among which are the Telecommunications Regulator Board, the National Agency for ICT, etc. The government has also set up a number of ICT training institutions including multimedia resource centres in some public secondary schools; a National Polytechnic, three Technology Institutes attached to state universities; an Advanced School of Posts and Telecommunication; and computer science departments in state universities. There equally exist numerous private initiatives in ICT training especially in private postsecondary institutions. There are genuine fears concerning the overlap in the duties of the myriad of ICT institutions in Cameroon and their cost implications. For example, there are two ICT strategies, the first by the Ministry of Post and Telecommunications (2005), and the other by the agency now charged with promoting ICT (ANTIC). ANTIC's strategy fails to offer any critique of the existing strategy to justify a new strategy.

The ICT regulatory environment witnessed a major shift following a decree liberalising the audiovisual sector in December 1990 and its texts of application in 2000. As concerns the telecom sector, several laws were enacted starting from 1998[4] with that reorganising the telecom sector and state withdrawal from mobile service provision. This equally marked the start of private internet service provision and the creation of the Telecom Regulatory Board. It was a major shift towards the liberalisation of the telecom sector. The regulatory environment has however failed to catch up with the rapid developments in ICT. The last texts, largely devoted to the telecommunications, were signed more than 8 years ago. There is no legal framework regulating the internet and information technology. For example, nothing is mentioned about cyber criminality; protection of ICT-related property rights, e-commerce, the legality of online transactions, etc.

3. Application Software Production and the Economy: A Review of the Literature

The literature on the impact of the production of application software on economic activity in developing countries is still scarce. However, a number of studies (Bamiro, 2007), narrate the experience of India and outline some of the factors responsible for the boom of its software industry. India is often cited as an example for other developing countries to emulate. This section thus focuses on how software production has transformed India's economy and the success factors.

The Indian IT-BPO (Business Process Outsourcing) sector is said to employ more than one and a half million people, with total aggregate revenue for the sector expected to exceed US$47.8 billion. According to NASSCOM (2009:5), "The IT-BPO industry in India has today become a growth engine for the economy, contributing substantially to increases in the GDP, urban employment and exports, to achieve the vision of a "young and resilient" India. The sector's share of total Indian exports (merchandise plus services) has increased from less than 4% in 1998 to almost 16% in 2008.

NASSCOM (2009) further estimates that the Indian IT-BPO industry will achieve revenues of US$71.7 billion in 2009, with the IT software and services industry accounting for US$60 billion of revenues. During this period, direct employment is expected to reach nearly 2.23 million, an addition of 226,000 employees, while indirect job creation is estimated to touch 8 million. The sector revenues will grow from 1.2% of GDP in 1998 to an estimated 5.8% in 2009. Net value-added to the economy is estimated at 3.5-4.1% for 2009.

The literature notes that developing countries have a comparative advantage in developing the software relative to the hardware industry, as software is a labour-intensive[5] and labour costs accounts for 70% to 80% of the total costs in software development (Bamiro, 2007: 20). There is a relatively low level of entry barriers and little effects of economies of scale. It is also argued that there are more opportunities for developing software in poor countries than hardware, as local needs for application software exist in almost any country.

A number of factors are accountable for the success of the software industry in India. First, the success of the software industry in India has largely been influenced by Government's focus on education that has helped create the large talent base (skilled and employable) from where the industry draws its workforce (NASSCOM, 2009). Over 3.5 million graduates and postgraduates are added annually to the talent base; no other country offers a similar mix and scale of human resources.

Second, an enabling policy environment following liberalisation, which allowed foreign capital and firms to enter the sector, facilitate knowledge and technology transfer and raising the level of competition thus preparing Indian firms for international competition. The creation of software parks and special economic zones was also very important for the Indian software industry. According to (NASSCOM, 2009: 9), the supportive policy environment created by the Government of India has played a key role in the rapid development of this sector. The role of government in the emergence of a strong ICT industry is corroborated by *Saraswati (2009: 1) who notes that* "... most serious observers now accept that the Indian state has played a major role in its nurture and growth over the past three decades".

The availability of power and quality of the telecommunications infrastructure (bandwidth and telecommunication penetration) were very important for the thriving of the software industry. Arora and Athreye (2002) note that the costs of power are the second highest expenditure for software firms and many of them are obliged to generate their own power. NASSCOM (2009) equally insists on how efficiencies in the provision of public services like power, local transportation, security, etc., are critical for the development of the Indian software industry.

Third, Indian firms strive to align themselves with international standards and are able to deliver high quality services and provide information security. As of December 2006, over 440 Indian companies had acquired quality certifications with 90 companies certified at SEI CMM Level 5 – higher than any other country in the world (Bamiro, 2007). To develop India as a trusted sourcing destination, they have put in place world class security standards, to ensure high levels of quality and service delivery (NASSCOM, 2009).

The hallmarks of the Indian IT-BPO industry in the words of NASSCOM (2009) are its strong fundamentals, a robust enabling environment, and enhanced value delivery capability. India enjoys a cost advantage of around 60-70% as compared to source markets.

The software industry holds great potentials for the developing countries of Africa if concerted efforts are made at evolving relevant policies and strategies to develop the industry, not only to meet the growing domestic markets but also the growing world market for software products and various IT-enabled services. The software industry is capable of creating jobs, increasing tax revenue, enhancing consumer spending, generating foreign exchange, supporting the establishment of other industries, thus contributing to economic growth.

4. Conceptual Framework[6] and Methodological Approach

This study uses as conceptual framework, a modified version of Porter's Five Force Model of industry analysis. In addition to the five competitive forces which are the major determinants of industry competition, the model is adapted to reflect the realities in most developing countries, where the government and other actors play important roles in shaping industry competition (Bamiro, 2007).

Figure 2: The Modified Porter's Model of Industry Analysis

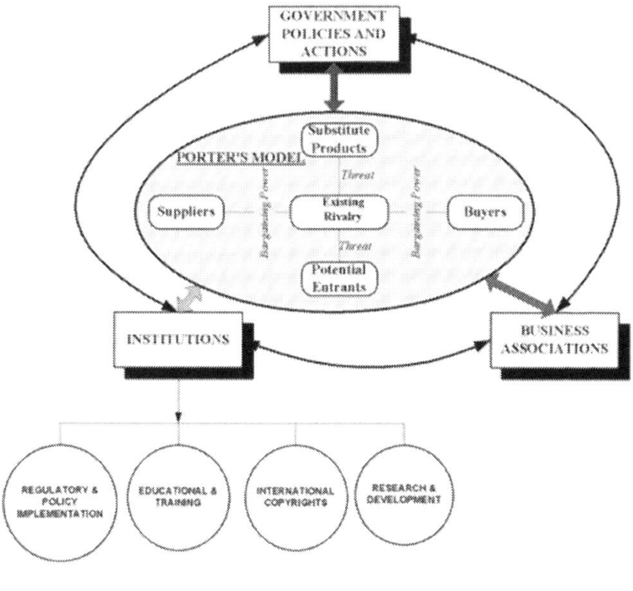

Source: Bamiro (2007)

Rivalry between existing competitors is the central competitive force. Several factors come into play when determining the intensity of rivalry among firms in the industry, including market size, industry growth and the demand/supply relationship. In highly competitive markets, firms monitor their rivals extensively and regularly, concerning price or product change and also try to bring in some innovations on their part. Another competitive threat comes from potential new entrants and depends on the nature of entry barriers. High profit margins and low entry barriers will attract new entrants into a market place. Barriers to the entry of new firms might arise from: existing firms benefiting from substantial economies of scale, product differentiation, high upfront capital requirements, high cost of switching to new products by firms, and access to distribution channels.

Substitutes to an existing product are a source of competitive threat as they introduce new technology or reduce the costs of producing the same product, without necessarily replacing it entirely. They however reduce the profits in industry by keeping prices down as competition is increased. The bargaining power of suppliers of branded software to local technical value-adding resellers (if we restrict ourselves to the software industry) also represents a threat. Porter suggested that under some conditions, such foreign suppliers can be more powerful and influence competition in the industry: a few foreign suppliers with no substitutes for their products, suppliers' prices form a large part of the total costs of the organization, and the foreign supplier can potentially undertake the value-adding process in the importing country. Under some conditions outlined by Porter, the bargaining power of buyers/customers might also represent a competitive threat: if there are a few and concentrated, if there is little or no product differentiation, and if it is possible for the buyer to integrate backward and take over the role of the organisation.

According to Bamiro (2007), the five-force model of industry analysis will be incomplete in Africa where the economic environment is highly influenced by government. Groups like business associations lobby government for an enabling business environment, protection against foreign competition, favours in contract awards, etc. Some institutions also play important roles like regulation, training, research and development, and those protecting intellectual property rights. Such institutions, components of the National Innovation System, enhance the capabilities of actors in the sector to face the challenges of competition and globalisation. Porter's five-force model of industry structure is thus modified by three key elements: government, business associations, and institutions as shown in Figure 3. It is within this general framework that we explore the market for application software used in the financial sector in Cameroon.

The methodological approach is broadly divided into two parts. First, we examine the application software market in Cameroon through the financial sector which is believed to be the most ICT-intensive sector in the country. We administered questionnaires to commercial banks and microfinance institutions (in the towns of Douala and Yaoundé), to identify the software used and the suppliers/developers of such software[7]. Secondly, we selected two local application software developers for case study. We interviewed them to know their range of products, required skills and competences, market share (including external market), turnover, and nature of operational and marketing constraints faced. The choice was on local developers, as we sort to assess the level of activity surrounding applications software industry in Cameroon.

5. The Market for Financial Sector Application Software in Cameroon

The financial sector is comprised of commercial banks, MFIs, informal financial institutions and other non-bank financial institutions like insurance companies, the Stock Exchange (DSX), etc. Financial institutions use application software for a wide variety of functions including banking operations, human resource management, payroll, accounting, customer relationship management, inventory management, etc. Some of the software packages are specific for front-office operations, while others perform back-office duties. This study is limited to application software used by commercial banks and MFIs.

Application Software in Commercial Banks in Cameroon

The survey results indicate that all commercial banks in Cameroon use foreign-made software. Most of these banks are affiliates of multinational corporations and naturally adopt the software used by the parent company. However, even locally-owned banks use foreign produced software. This might be an indication of the absence of a viable locally-made substitute or lack of trust of local software developers by local banks. As indicated on Table 4, DELTA Bank is the dominate software, used by seven of the twelve commercial banks (i.e. 58 per cent). DELTA Bank is produced by 'Delta Informatique', a French company, and all local subsidiaries of French banks use this software, as well as all Cameroonian-owned banks, some of which have very close relations with French banks or nationals. Those not using DELTA Bank are largely subsidiaries of American, British and Nigerian banks.

What is the competitive nature of the commercial bank software market in Cameroon using Porter's modified model? Most of the banks are subsidiaries of multinational banks and consequently use software from the parent bank. The market for commercial banking software therefore extends beyond the boundaries of Cameroon, as none of the software is produced locally and most of the buying decisions are taken out of Cameroon. At the international level, there are many banking software suppliers and also many buyers. This implies rivalry between existing firms and the existence of substitute software and potential entrants in the market.

Table 4: Software used by Commercial Banks in Cameroon

Name of Commercial Banks	Software used
Afriland First Bank	DELTA Bank
Banque Atlantique	DELTA Bank
Banque International du Cameroun pour l'Epargne et le Crédit – BICEC	DELTA Bank
CITI Bank	FLEXCUBE
Commercial Bank of Cameroon	DELTA Bank
ECO Bank	GLOBUS
National Financial Credit Bank	DELTA Bank
Crédit Agricole - Société Camerounaise des Banques (CA - SCB)	DELTA Bank
Société Générale des Banques au Cameroun – SGBC	DELTA Bank
Standard Chartered Bank	Bank Master
Union Bank of Africa	FINACLE
Union Bank of Cameroon/OCEANIC	FINACLE

Source: Survey of Commercial Banks by authors in July 2009

Such a competitive market reduces the bargaining powers of both suppliers and buyers, and they can only retain some bargaining power if they pool together in the form of business associations. The high cost of migrating from one software package to another might be used by some suppliers to retain some bargaining power especially when buyers move to update software. More generally, the competitive nature of the commercial banking software industry constitutes an entry barrier to potential newcomers in the market, especially local developers in Cameroon. Given the international nature of this market, local business associations, the government or local institutions (like regulatory bodies and educational/training institutions) should not have any influence on its competitiveness.

Application Software in Microfinance Institutions in Cameroon

Apart from commercial banks, MFIs are also important users of application software in Cameroon. One significant feature of MFIs in Cameroon with important implications for the application software market is their tendency to group into networks. According to La Microfinance (2008), of the 490

MFIs in Cameroon by end of 2006, 309 were grouped into 7 networks[8]. The largest of them, CamCCUL (Cameroon Cooperative Credit Union League) grouped 177 MFIs. These networks influence the choice of software used by members. For example, within the CamCCUL network, only three software packages are currently used (Alpha, Flextra and Gescoopec), while the MFIs in the MC² network all use ALT MicroFi.

ALT MicroFi is developed by Afrik Brain which belongs to same holding company (Afriland Group) as Afriland First Bank. Afriland First Bank is the promoter of all the 57 MFIs in the MC² network (as of end 2006) and MUFFA[9] (also using ALT MicroFi). Similarly, all MFIs in the CamCCUL network buy their software through CamCCUL, as payments in instalments, including the maintenance fee are made through CamCCUL. The most frequently used software packages in the microfinance sector are not the most competitive (in terms of performance and price), but are largely attributed to the networks or business associations to which the MFIs belong.

Aside from the MFIs grouped in networks, we successfully surveyed 23[10] other MFIs in the towns of Douala and Yaoundé. The survey identified other software used by MFIs: Eloge Bank, Softlink, La Logique, Turbo Bank, NetSoft and Real Bank. Of all these, only Real Bank is foreign; imported from Kenya but developed in India by Craft Silicon. Eloge Bank is the most popular software, used by 9 out of 23 MFIs surveyed and not belonging to the seven MFI networks mentioned above. Eloge Bank is developed by ICCSoft (one of our case studies) and designed specifically to meet the needs of MFIs. Flextra, Turbo Bank and Netsoft were the next most used software with 3 users each from the 23 MFIs surveyed.

The market for software used by MFIs is different as software is largely home-produced and home-consumed. We identified a substantial number of software in the market, and the large number of MFIs in the country points to a high level of competition in the market i.e. with the existence of many buyers and suppliers and also with most of the software products in the market being direct substitutes for one another. The level of rivalry in the market is thus expected to be high and should constitute an obstacle to new entrants, even though upfront capital requirements are not very high in the software industry. One will also expect the bargaining power of both buyers and suppliers to be low in a competitive market, though suppliers will have an edge given the high cost of switching to new software.

Limiting ourselves to Porter's original model leads to the conclusion that the microfinance application software market in Cameroon is very competitive. However, if the influence of business associations or the MFI networks that characterise the microfinance landscape in Cameroon, is taken into consideration, a different picture of the market emerge. As indicated above, MFI networks influence the choice of software used by members and consequently reduce the level of competition in the microfinance software market in Cameroon. For the moment, the government plays no direct role in the market, as it neither supports software development nor regulates the market, even at the level of fighting piracy. The expectation is that local developers might make their way into the commercial bank software market as they gain more experience and as some of the big MFIs mature into commercial banks, while maintaining their grip on the microfinance software market where they appear to be satisfying the needs of the customers.

Case Studies of Software Developers in Cameroon

Azerty Informatique et Multimedia

This firm was created in 1998 by a group of young Cameroonians, mostly graduates from the Faculty of Science of the University of Yaoundé I, who had worked for other software firms upon graduation. It specialises in the development of application software, but also undertakes other ICT activities like editing of web pages, network creation and maintenance, ICT training, creation and editing of films for enterprises and advises firms on the choice of ICT infrastructure. The Department of Information Technology (one of five in the firm) has the responsibility of developing application software. At the time of the interview, Azerty employed 13 permanent workers (including 10 ICT specialists) and 4 consultants.

Table 5 lists the main application software (and their functions) developed by Azerty Informatique since creation. Cyber2000 was the first application software developed more than 10 years ago, and is still widely used in the country today. This software automates and facilitates the operations of Cybercafés but can also be used to sell and keep records of any type of good or service; thanks to its ability to manage several cash desks and many accounts. A number of firms including hotels, transport, etc are using standardised software developed by Azerty and several other software applications are developed to meet specific needs of clients.

The firm has been encountering a number of problems and equally has many grievances towards ICT authorities in Cameroon. The first category of problems has to do with software development. The cost of acquiring licences for software development platforms (like Fortran, Cobol, C++, Visual Basic, SQL, Java, etc.) is very high. There is also the lack of qualified and experienced developers as those who join the profession after graduation quickly quit because of less attractive salaries.

Table 5: The Main Software Produced and Marketed by Azerty Informatique

Software Name	Functions
OManager, EmptyEnd and EasyManager	Business management for Partners (licensed dealers) of each of the three mobile phone providers: MTN, Orange and CAMTEL
CheckPoint /Hotel2000	Hotel Management and Accounting
TravelSoft	Management software for firms in the inter-urban transport business
SunrisePlus	Business management
Cyber2000	The first software for the Management of Cybercafés in Cameroon
Paraphama	Management of pharmacies/Medicine stores
PaP2000	Management of fashion shops

Sources: Website of Azerty Informatique; and Interview with the General Manager in July 2009.

The other set of problems is related to the marketing of software. The rate of software piracy in Cameroon, as mentioned before is endemic. Azerty has successfully protected its software from piracy, but pirated copies of rival imported software are largely and cheaply available on the market. Some of the pirated software is sold for less than one-twentieth, the price of similar software marketed by Azerty. Another marketing problem has to do with the inability or unwillingness of Cameroonians to computerise their operations – some for fear of transparency. The firm has not sold a single copy of PaP2000 and ParaPhama since their development. The award of government software contracts is also deemed non-transparent and most of the contracts are awarded to foreigners at the expense of local developers. The government is, according to Azerty, failing in its duty to patronise software production in Cameroon and secure/protect existing software.

The management of Azerty affirms that their success has been built on three points: speed, stability and security of their products. Software is sold to customers according to their financial resources and the firm helps customers to choose the format most appropriate for their needs. Azerty has been successfully exporting software to neighbouring countries in the sub-region like Gabon, Congo, Chad and the Central African Republic. Software packages exported include Cyber2000, Omanager and especially CheckPoint.

The International Computer Centre (ICCSOFT)

ICCSOFT is born from the disaggregation of the parent company ICCNET (now ICCNET Holdings) into several ICT operations. ICCSOFT specialises in application software development, web applications and hosting of websites and became autonomous in January 2003, inherited seven years of experience in software development from ICCNET Holding. ICCSOFT employs 15 workers, including 10 software developers; among whom is their lady General Manager.

The main product of ICCSOFT is Eloge Bank, designed specifically to meet the needs of MFIs. La Microfinance[11] offers a favourable review of this software, describing it as software for quick, simple and efficient management of an unlimited number of bank accounts that offers all the tools necessary for optimal management of bank accounts and is completely secure. A web version of Eloge Bank is available, allowing customers to connect to accounts through SMS. ICCSOFT has a technical team in many parts of the country to support users of the software. It has also created Eloge Forum - a discussion forum on its products for users to share their experiences and find quick solutions to problems with the software.

A substantial number of microfinance institutions in Cameroon are reportedly using Eloge Bank for their transactions, among which are some of the most successful ones like COFINEST and La Régionale. Eloge Bank sells for €3,000 or CFAF2 million, but the firm is prepared to sell below this price in the local market. ICCSOFT reports on its website that the software is already making inroads into neighbouring countries like Chad, Central African Republic and Congo.

Their main difficulty in developing software is lack of experienced and devoted developers. Those recruited are impatient and quit before gaining the required experience. Concerning marketing, they complain of software piracy, even though theirs is protected. The most pirated packages in the market are Saari and Ciel. They also decry what they term the 'mentality' of some Cameroonian managers who prefer expensive foreign software, which might not even be as good as that produced locally. The award of government software development contracts to foreigners is seen as government failure to promote local development initiatives. They wished government could reduce taxes and offer them financial assistance given the number of Cameroonians they employ, despite the sluggish application software market.

6. Conclusion

The government of Cameroon is highly committed to vulgarise ICT in the country, with the aim of making ICT a tool of socioeconomic development, poverty reduction and good governance. However, the focus has been on promoting ICT use and little or nothing on ICT production. The two ICT strategies elaborated in the country so far, fail to chart a clear course for promoting application software production, despite the numerous opportunities. Furthermore, there are too many government institutions responsible for fostering ICT in the country with the risk of overlapping responsibilities and the ICT regulatory framework is not catching up with the rapid evolution in the ICT sector.

Though, both of the financial sector, the market for commercial bank application software is distinct from that for MFIs. The market for commercial bank software is international and consequently beyond the influence of the local environment. Most of the banks are subsidiaries of multinational banks and understandably use same application software as parent bank. The market for microfinance software on the other hand, is local and the level of competition is largely influenced by MFI networks/associations. The expectation is that local developers will eventually break into the market for commercial banking software as they gain more experience and as some of the big MFIs mature into commercial banks, while maintaining their grip on the microfinance software market.

The case studies of some local application software developers reveal that there are a few well-structured and very ambitious firms involved in software development in Cameroon. They are not all making good business but are able to break-even and stay afloat. Their main problems are related to piracy, unavailability of experienced and devoted workers/developers, lack of demand for their products, absence of government patronage and the award of most government software development contracts to foreigners, when local firms are capable of doing the job. It is also important to underscore that many more software developers do not even have an office, not to talk of websites and operate almost clandestinely for fear of taxes.

A number of policy implications follow from our examination of the application software industry in Cameroon. There is need to pay more attention to application software production (and not only ICT use) and include this in the ICT sector strategy as there is a niche for local developers in a largely unexplored market. A tax break, government patronage and fight against software piracy (not by the Ministry of Culture as is the case now) could be helpful. There is also a need to streamline ICT institutions i.e. redefining the responsibility of the various institutions and closing down those that have become obsolete. The ICT development authority (ANTIC) should set up a special service devoted entirely to the promotion of local software development industry. A new regulatory framework is also required to meet up with rapid evolution in ICT in Cameroon and the world.

Notes

1 ANTIC stands for Agence Nationale des Technologies de l'Information et de la Communication.

2 Our translation of part of the President's speech made in the French language on 10 February 2010.

3 According to Nana Nzepa and Keutchankeu (2006: 72) there were 36 providers by April 2006.

4 Law No 98/014 of 14th July 1998.

5 The software industry might be labour-intensive, but the quality of labour required is an important issue. Some developing countries, especially in Sub-Saharan Africa might have the labour but the level of development of the educational system prevents them from reaping the advantages offered by the software industry.

6 We present only the highlights of the Modified Porter's Model of Industry Analysis in this study. Interested readers are referred to Bamiro (2007).

7 Questions related to financial issues like expenditure on software were systematic ignored by the respondents.

8 The 7 MFI networks are: CamCCUL, le Réseau de l'Association des Caisses Villageoises d'Epargne et de Crédit Autogérées du Centre, le Réseau de l'Union des Caisses Villageoises d'Epargne et de Crédit Autogérées du Grand Nord, the Network of Noweforch Savings and Loans Cooperative Societies of North West, le Réseau des Caisses Mutuelles d'Epargne et de Crédits du Grand Nord, and Le Mutuelle Communautaire de Croissance (MC2).

9 MUFFA is acronym for 'Mutuelle Financière des Femmes Africaines' (Mutual Fund for African Women).

10 Some of the 50 MFIs targeted refused to fill the questionnaires and the information provided by others was inadequate. The main reason being that, most of them use pirated software and consequently perceived the exercise as an attempt to unmask them. Some other institutions saw the interviewers as potential rivals seeking for the appropriate software to use.

11 La Microfinance is the French equivalence of Microfinance Gateway, the portal for MFIs: http://www.lamicrofinance.org/resource_centers/systemeinformation/article/23636/

References

Agence de Régulation des Télécommunications (ART) (2006), Information Trimestrielle sur le Marche des Télécommunications, *Bulletin d'Information Trimestrielle*, No. 3. Yaoundé

Agence de Régulation des Télécommunications (ART) (2008), *Nouvelles de l'ART*, No 21, April-May, Yaounde.

Agence Nationale des Technologies de l'Information et de la Communication (AN-

TIC) (2007), *Stratégie Nationale de Développement des Technologies de l'Information et de la Communication*, ANTIC, Yaounde.

Arora, A. and Athreye, S. (2002), 'The Software Industry and India's Economic Development,' in *Information Economics and Policy*, 14(2): 253–273.

Bamiro, F. (2007), 'The Vision and Challenges of ICT Production in Africa: Software Production and Services', *Framework paper for the AERC project - The Impact of ICTs on Economic Development and Transformation in Africa*, Nairobi, Kenya. See: www.aercafrica.org/publications

Cameroun (Undated), 'Révision du DSRP: Document de travail', Yaoundé: Ministry of Planning and Economic Development.

La Microfinance (2008) 'Profil Pays du Cameroun,' Available at: http://www.lamicrofinance.org/resource_centers/profilcameroun (Accessed on: 03 September 2009).

Ministry of Posts and Telecommunications (2005), *Sector Strategy Document for Telecommunications and Information and Communication Technologies: 2005-2015*, Yaoundé: MINPT.

Nana-Nzepa, O. (2003), *Building the Future: Civil Society's Contribution towards the Emergence of the Information Society in Cameron*, Johannesburg: APC Africa Rights, Available at: http://africa.rights.apc.org/research_reports/cameroon.pdf

Nana-Nzépa, O. and Tankeu-Keutchankeu, R. (2006), 'SME-Sector: Cameroon', in *African e-Index: Towards an SME e-Access and Usage across 14 African countries*, Research ICT Africa. Available at: www.researchICTafrica.net.

National Association of Software and Services Companies (NASSCOM) (2004), *Nasscom Strategic Review 2009*, New Delhi: NASSCOM

Ngae, D. and Nkolo, E. (2007), 'Cameroun', *Atelier de Planification Stratégique pour la fourniture de services aux Zone Rurales et éloignées d'Afrique Centrale*, Bangui. Available at: www.itu.int/ITU

Nour, S. O. M. (2002), 'ICT Opportunities and Challenges for Development in the Arab World', *UN WIDER Discussion Paper*, No. 2002/83, United Nations University, Maastricht, Netherlands.

Oyeyinka, B. (2007), 'ICT and Industrial Development, Transformation and Employment Generation', *Framework paper for the AERC project: The Impact of ICTs on Economic Development and Transformation in Africa*, Nairobi, Kenya.

Pohjola, M. (2001), *IT, Productivity and Economic Growth*, London: Oxford University Press

Saraswati, J. (2009), 'The Debate on the Developmental State: Evidence from the Indian Software Industry', *Development View Point No. 31*, Centre for Development Policy and Research, School of Oriental and African Studies, University of London.

Wangwe, S. (2007), 'Evolution, Status and Impact of ICT on Economic Development and Transformation in Africa: An Overview', *Framework paper for AERC project: The Impact of ICTs on Economic Development and Transformation in Africa*, Nairobi, Kenya.

World Bank (2008), *ICT at a Glance: Cameroon*, Washington, D.C.: World Bank.

Chapter 11

Link between ICT Investments, Growth and Jobs Creation in Senegal

Latif Dramani & Oumy Laye

1. Introduction

The government of Senegal has adopted the Accelerated Growth Strategy which uses the wealth creation of the PRSP in order to double the national income by 2015. This program will rely on a strong and sustained 7% to 8% per year until 2015. The Accelerated Growth Strategy (CAS) has identified the ICT sector as the main engine of economic growth in Senegal, as one of the sectors with high potential.

Classical studies or specific research on the analysis of the impact of investments in ICT on the living conditions of households are few and often unsuited to the context of developing countries, given the methodologies used. Our analysis attempts to assess the impact of sectoral policies on economic growth using a model of the multiplier of social accounting matrix (SAM).

With the development of new technologies of information and communication technologies (ICTs) during the 80s and 90s, relevant improvements in productivity were expected. However, empirical studies have generally found no relevant productivity improvements associated with investment in ICT.

The productivity paradox, still called Solow paradox, is that ICT development was supposed to have a positive effect on productivity and thus boost growth. In fact, because of their speed, efficiency and lower cost, intensive use of ICT as an input would necessarily lead to increased production for the same amount of work. In Senegal, the contribution of telecommunications to growth is quite significant and represents 6% of GDP in 2005. Given the empirical results showed that the development of ICT had no significant impact on productivity, it is essential to ask whether the development of ICT in Senegal is associated with improved productivity and hence growth economy.

The introduction of new information technologies and communication in a process of production has positive and negative effects on employment. ICTs call for a reorganization of work that can result in creation as a job destruction. In these circumstances, it is important to consider the development impact of ICT investment on employment in Senegal.

The concerns raised above include the overall issue of this work is summarized in the following question: massive investments in ICTs in Senegal are an obstacle or a catalyst for growth and employment?

2. Stylized Facts

Profile different areas of investments in Senegal

Several economic agents invest: households, government and businesses. At company level, three areas are sometimes distinguished: non-financial corporations, unincorporated businesses and financial companies. Non-financial companies also include companies that are majority owned by the government. During the period 2000-2005, the average GFCF in Senegal amounted to 610 billion CFA francs, or about 21% of GDP.

Table 1: Investment by Agent (billions CFA)

	1980-84	1985-89	1990-94	1995-1999	2000-2005
Households	100	99	130	142	309
Financial Firms	53	89	128	302	355
Public administration	31	51	65	138	192
GDP growth	2.9%	2.4%	2.1%	4.1%	4.7%

Source: Author estimations from ANSD (national accounts database, Method of estimation: mean average for 5 years)

The total investment is traditionally divided into three distinct sets: the government investment, investment by households and investment companies and financial institutions, individual entrepreneurs or companies, which is traditionally reserved for the name of productive investment because that is investment-oriented production of goods and services.

Government

Public investment has remained low during the periods 80-84, 85-89 and 90-94. It has accelerated dramatically from early 1996 to an average of 138 billion CFA francs as against 65 billion over the period 1990-1994. This increase has continued during the period 2000-2005 as a result of increased Consolidated Investment Budget (BCI). This increase can be explained by an increased attention to land transport (contributing 39% of the increase in the BCI during this period), followed by administrative facilities (23.4%) and health (13.5%). It must be noted that much public investment is financed by resources from Foreign Direct Investment.

Business

The total value of investment companies and financial institutions during 1980-84 and 1985-89 were 53 billion and 89 billion CFA francs respectively.

After the devaluation, the volume of GFCF companies had averaged 302 billion CFA francs against 128 billion CFA francs in the period 1990-1994. Between 2000 and 2005, the volume of productive investment had been particularly dynamic reaching 342 billion CFA francs. In 2005, the changing composition of investment companies brought out the development of market services, which were increased from 57% of total investment in 2000 (68 billion CFA francs) to nearly 65% in 2005, while industrial investment in currency regressed (from 42.1 to 35.3 billion CFA). The investment activities of services mainly merchants in 2005 represented more than three times that of the industry. This corresponds to a fundamental trend of industrialized economies, following the rise of the tertiary sector. It also demonstrates the restrictive nature (or partial) investigation focused on industrial investment that does not describe or recount a share of less and less of the investment (which remains crucial to the competitiveness of the Senegalese economy).

Today, the investment is mainly attributable to the services sector, and transportation. In proportion to their turnover, small and medium-sized firms invest more than large firms. Firms with fewer than 20 employees, representing 21.8% of total turnover of Senegalese companies, invested over 31% of the total effort of business investment. A policy to support the investment must take into account the important role of SMEs.

Households

The household sector consists of the informal sector, housing and other household investments constituting the largest share of their investment. During the periods 1980-1984 and 1985-89, household investment amounted respectively to 100 and 99 billion on average. From the 90s, it increased, amounting to 142 billion on average during the period 95-99. With the change of regime in 2000, a renewed dynamism of the investment is noted. It averaged 343 billion during the period 2000-2005.

Contribution to growth sectors in Senegal

The contribution of Posts and Telecommunications to the growth of the Senegalese economy has been increasing, apart from 1991-1995, when it registered a -0.01% contribution to GDP. During the period 2001-2005, its contribution to GDP reached 0.94%. It is followed by the trade sector (traditional bearer of growth) and the construction sector.

Table 2: Contribution of the Sectors to GDP Growth in Senegal (%)

	80-85	86-90	91-95	96-2000	2001-2005
Total Value added	2.5	2.0	1.8	3.5	3.7
Primary sector	0.9	0.1	0.2	0.5	-0.1
Secondary sector	0.5	0.6	0.7	0.9	0.8
Tertiary Sector	1.1	1.3	0.9	2.2	3.0
GDP	2.95	2.39	2.05	4.20	4.22
Food firms	0.17	0.21	0.08	0.12	0.00
Energy (Water, Electricity, Gas)	0.07	0.00	0.09	0.13	0.14
Buildings firms	0.08	0.09	0.14	0.25	0.41
Trade firms	0.53	0.40	0.42	0.74	0.56
Repairing services	0.02	0.02	0.02	0.02	0.03
Restaurant and Hotel Services	0.04	0.05	0.01	0.02	0.01
Transports	0.14	0.04	0.05	0.03	0.23
ICT	0.03	0.20	-0.01	0.42	0.94
Financial services	0.13	0.09	0.08	0.13	0.20
Leasing Activities	0.18	0.18	0.19	0.23	0.18
Services for firms	0.09	0.09	0.10	0.27	0.36
Administration	-0.10	0.19	0.06	0.29	0.45

Source: Author estimations from ANSD (national accounts database)

The development of the telecom sector is linked to several reasons: (i) the restructuring of Sonatel, which has been privatized to break records of profits; (ii) trade policy more responsive and more aggressive; (iii) the popularization of mobile phones whose prices have sharply lowered, and this opportunity became available to a wider audience; (iv) the effect of competition, allowing more choices for players; and (v) improving the supply of internet connection.

The table above shows that despite a relatively low weight (less than 6% of GDP in 2005) compared to other industries such as trade (16% of GDP in 2005), agriculture (8% of GDP 2005), the contribution of telecommunications to growth is very important. It's practically the telecom which drove growth in Senegal for five years. This indicates that the structure of the economy in terms of ICT use is changing.

According to the survey of Poverty Monitoring (ESPS), over 20% of poor households have a cellphone, 6% fixed phones and over 20% of poor households have a television.

Application of ICT in Senegal: Link between poverty and ICT spending

Table 3: Poverty and Households Expense on ICT

Expenses (CFA)	Rich	Rank	Poor	Rank	Total	Rank
Health expenses	96 275.5	3	39 955.4	3	136230.9	3
Education expenses	57 343.5	6	15 377.9	7	72721.3	6
Transport expenses	111 835.3	2	40 707.0	2	152542.3	2
Repairing & building modifications	18 674.6	8	5 580.7	8	24255.2	8
Expenses on water	50 692.9	7	18 878.2	5	69571.2	7
Expense on electricity	83 234.9	4	18 273.5	6	101508.4	4
Expense on ICT	74 365.5	5	21 256.3	4	95621.7	5
Households expenses	208 293.4	1	58 407.9	1	266701.3	1

Source: Author estimations from ESPS (Third national household survey in Senegal) The value are the mean average expense of the households on each item of good.

Analysis of Table 3 shows that, on average, households spend more in the teleservices products (95.6 billion CFA francs) in water consumption (69.57 billion CFA), or school expenses (72 billion). However, in poor households it is observed that the resources allocated to spending on Teleservices (21.256 billion CFA) were higher than those allocated to education spending (CFA 15.38 billion), electricity (CFA 18.27 billion) and water (CFA 18.88 billion). Spending on online services is placed in order of importance in the 4th position among poor households, while wealthy households in the teleservices come in 5th position.

The average expenditure of poor households on telephone is 25% of total expenditure, exceeding even the cost of water and electricity taken separately. Expenditure on goods and services not otherwise specified are estimated at 84.449 billion or 4.2% of household expenditure. They consist of personal care expenses, personal effects not reported on the topics mentioned above, communication costs and other services not listed elsewhere.

The most important part of spending on goods and other services results to the costs of communication with 32.106 billion (38.0%), following expenditure on personal care and chattels not reported elsewhere.

This structure hides an uneven distribution of expenditure according to different subgroups. Spending on communication is related to the degree of urbanization. Households spending on communication in Dakar amounted to 48.9%, in other urban centres 39.9% and among rural households 13.8%. Spending on communication increased between 1994/1995 and 2001/2002. They have doubled over the period from 19.2% to 38.0% at the expense of spending on personal care and personal effects not otherwise specified (36.9 and 35.5% respectively in 1994 and 1995).

3. Selective review of literature

Impact of ICTs on productivity and growth

The link between information technology and communication (ICT) and productivity growth has long been discussed over the three decades. During the 1980s and early 1990s, empirical research generally did not find improvements (Roach, 1989). At that time, economists expected a positive effect of ICT investments on productivity. Indeed, intensive use of ICT as an input should lead to increased production for the same amount of work.

From the years 1994-1995, several economists (including Oliner and Sichel, 2000 and Jorgenson and Stiroh, 2000) can identify a positive contribution of IT investments to grow the economy.

The negative results of ICT on growth even called "Solow paradox" could be explained by the fact that it takes large changes in capital stock for the impacts of these changes are significant and can be measured on production. Brynjolfsson and Hitt (1993) gives several reasons why the Solow paradox. He cited technical problems, the lag between ICT investment and profits, the distribution of gains among firms with intensive use of ICT and those who invest little, and that lax management at the use of new technologies.

More recently, with the availability of new data and new methodologies, the impact of ICT on productivity has been measured. Among the economists who first established the link between labour productivity and new information technologies are Brynjolfsson and Hitt (1996). The results of their estimates showed that the firm fixed-effect accounted for almost half of productivity gains related to information technology expenditures and information technology had a significant elasticity of output.

In line with the macroeconomic studies conducted by Jorgenson and Stiroh (2000) on the impact of ICT on growth in U.S. productivity, other empirical studies have shown a strong impact of ICT on economic growth and productivity (Oliner and Sichel, 1994; Lehr and Licthemberg, 1999; Sichel, 1997). To be productive, the installation of ICT requires complementary investments in communications equipment and software, and other elements such as human capital and organizational factors (Dedrick et al., 2003; Pilat, 2004). Capital ICT function as the implementation of further innovations in many activities of the company with a clear advantage for companies that undertake projects in research and development. This is more accentuated in the presence of high rigidity market that prevents successful companies to emerge, reducing the incentives to investment in technology. According Bassanini and Scarpetta (2002), institutional factors have made Europe an economic area only dynamic and partly explain why ICT has contributed to economic growth to a lesser extent than in the USA.

In seeking the sources of recurrence of the U.S. industry, Stiroh (2002b) and Nordhaus (2005) observe that comes entirely from sectors that produce and use digital assets intensively. Concerning the USA, O'Mahony and Van Ark (2003) noted a delay in service sectors such as finance, wholesale and retail industries where new types of technology have facilitated the re-organization in the last radical decade (McGuckin et al., 2004)).

Econometric estimates the link between the capital increase of technology and increased labor productivity in industry is mixed. According to Stiroh (2002a), industrial sectors Americans have not taken a particular advantage of the whole technology. On the other hand, O'Mahony and Vecchi (2005) found a positive effect for all industries American market, but not for the United Kingdom. For the latter, however, is a favourable indication of Oulton and Srinivasan (2005).

An unorthodox attempt to assess the existence of externalities of ICT knowledge is carried out by Dutta and Otsuka (2004), using data patent application for a small group of nations. Since the flow of new knowledge is strongly and positively correlated with stocks of patents used by leading industries, ICT patents are used as proxy of knowledge in the context of a production function. However, contrary to the requirements of new theories of growth, GDP is not affected by the input of knowledge, which may be related to the very short period considered in this study. In the same vein, the work of Becchetti and Adriani (2005) consider ICT as a vehicle for the dissemination of knowledge. In this regard, the use of new technology is emerging as a crucial additional factor in explaining income differentials across countries.

Fuss and Waverman (2005) showed that the externality of ICT was the main contributor to productivity for most OECD countries in the late 1990s. Following this line of discussion, it is reasonable to believe that the externality of productivity could be even greater than expected because it does not benefit from the link with the capital of information technologies and communication of partner countries commercial.

Lee and Guo (2004) explain the peak ICT investment abroad as a determinant of national growth of total factor productivity, yielding a robust evidence for spillovers from richer countries to less developed countries.

Nevertheless, many economists point out that we should remain cautious, in fact, if ICTs are essential for businesses, should not be seen as a factor that could in itself spur growth in all sectors of activity.

Industrial Transformation and Generation of Employment

With the development and adoption of ICT in various economic activities, the effects on employment have been controversial.

Some authors consider that ICT leads to mass unemployment, others believe that ICT is generating jobs and cause the creation of new industries and services (Talero and Gaudette, 1995). For some authors, the adoption of these technologies could result in reduced levels of activity, especially that of workers semi - and unskilled workers. This view has stressed the device programmability of ICTs and their potential in managing multiple tasks with a simple tool of ICT. Therefore, the perception emerged that the adoption of ICT has had a negative effect on employment. People still believe that the ICT tools replace certain categories of workers and lead to a significant rise in levels of unemployment.

Rada (1982) and Kuwahara (1984) found that the adoption of information technology and communication (ICT) in business from a static perspective leads to the creation of new jobs in some process and loss of manufacturing jobs in other activities. Freeman and Soete (1985) found no negative relationship between employment and the use of ICT. Their conclusions are based on several other studies (Leontief and Duchin, 1983; Kuwahara, 1984: Lawrence, 1984) conducted in developed countries (USA, Canada and Japan).

Peitchinis (1984) studied the employment effects of the introduction of computer hardware and office in a number of Canadian manufacturing industries ranging from food to oil companies. The author rejected the prediction of mass unemployment due to the adoption of ICT. Several other studies (Rada, 1982, James, 1994; Rahim and Pennings, 1987) have proposed the possibility of structural change in employment. Rada (1982) found evidence of a reduction in employment level of staff supervision. At the same time, the tools of information technology require a skilled workforce for the implementation and use of database systems Microelectronics (Ayres, 1991, James, 1994).

The developing countries have experienced similar effects on the use of information technology and communication. Acero (1995) reported changes in the structure of employment in a study of the Brazilian textile industry. Acero has found that industrial automation and technology organization are seen as contributing to increased levels of employment in the technical and administrative category, while the number of occupational categories and intensive tasks of labour work is decreasing with the introduction of ICT. Sim and Yong (1995) found similar results in their study on the Malaysian telecommunications industry. Doms et al. (1997) examined the correlation between the share of workers in non-production with the use of advanced technology. The authors found a positive correlation between two variables satisfying the complementarity of the intensity of competence in the use of technology.

4. Methodology

Methodology for development of the SAM and simulation

The ultimate goal is to construct a simulation model, it is important initially to construct a Social Accounting Matrix (SAM) or use an already constructed matrix that constitutes the empirical basis of computable general equilibrium models. The SAM is a static table that describes all trade flows between different economic agents. It is based on the principle of balance between jobs and resources. The SAM is in fact a generalization of matrix input-output "which trace the inter-industry trade (see Table 1). It is presented in tabular form in line with the column resources expenditure. It describes the national economy by organizing in different accounts that can be synthesized as follows using Table 1:

The account "product" describes the process of production (input-output table) and how the output is used by other industries or sectors (intermediate and final consumption, exports).

The account of factors (labour and capital) which helps to highlight the remuneration of factors between different economic agents; the account of agents (households, businesses, government) that trace their resources and their jobs; the capital account or accumulation to verify that all of the savings used to finance investment; the Rest of world (ROW) which traces the receipts (exports) Rdm and expenditures (imports) and the difference between the two is either saving or foreign borrowing outside the country.

The SAM can summarize in a single table all transactions between economic agents and are, in fact, a synthesis of TES and TEE. They were developed from the seventies to the specific needs of policy development, particularly to address issues related to employment and income distribution (e.g. poverty).

The major advantage of SAM is to represent a summary of all the interrelationships between income distribution and the structure of production. For this, the SAM based on detailed production accounts by industry, an account of factors of production and accounts for institutional sectors. This representation can lead to detailed classifications of factors of production and accounts for institutional sectors.

The simulation model

In a series of seminal contributions, Pyatt and Round (1979), Defourny and Thorbecke (1984) and Round (1985) each method developed for a large empirical economic analysis is based on social accounting matrix (SAM). They have been extensively used to examine the process generating income in many countries. One can also note that the aspect of unification of this work was a priority on the concept (quantity-driven determination revenue), the main side's underlying accounting system.

The SAM has been used for example to study: (i) strategies for growth in developing economies, Pyatt and Round (1985), (ii) income distribution, Pyatt and Roe (1977) and Adelman and Robinson (1978), and redistribution, Roland-Holst and Sancho (1992), (iii) tax policy at the national or regional, St Hilaire and Whalley (1983, 1987), and (iv) the breakdown of activity multipliers which threw light on the circuits with the circular flow of income, Stone (1981), Pyatt and Round (1979), Defourny and Thorbecke (1984), and Robinson and Roland-Holst (1988).

From the SAM multipliers can be derived with the same properties as those of input-output model. Using these multipliers, we can, for example, measuring the impact of economic policies on income distribution and resource allocation. This is the case in this mission, where it is about tying the performance of sectors and the strategy against poverty.

Using the model of the SAM multiplier which implicitly assumes exogeneity of government account and endogeneity for other accounts, it is possible to derive the impact of a change in the level of government spending on all components of the economy. In particular, the interest will be focused here on the capital expenditure made by the government during the base year (2005).

The Walrasian general equilibrium is the theoretical framework that constitutes the bedrock of calculations to measure the impact of a surge in investment on the economic structure in general and on each account of the SAM in particular. This balance describes the condition of a market economy where all markets are in equilibrium classical. The impact of the input of capital spending is apprehended on the overall output consists of aggregate income and income from each account in the SAM.

The SAM multipliers

The SAM allows to disaggregate the various accounts in the above-mentioned accounts into endogenous and exogenous accounts. For purposes of simulation, we assume that the column vector of coefficients is constant. The disaggregation of the column is as follows:

Accounts	**Endogenous accounts**	**Sum of exogenous accounts**	**Total**
Endogenous accounts (n)	AX	D	X
Exogenous accounts (m)	BX	L	
Total	X		

The vectors are defined as follows:

X = vector of income or expenditure of endogenous accounts;

D = vector sum of spending by exogenous accounts;

L = column vector of exogenous income accounts;

A = matrix (nxn) coefficient of endogenous accounts;

B = matrix (mxn) coefficient with the accounts as exogenous and endogenous accounts lines as columns.

The measures of the impact of policy shocks on the system are made by reasoning in terms of variation.

The multiplier matrix $(I - A)^{-1}$;

The vector of shocks ΔD ;

The vector of impacts $\Delta X = (I - A)^{-1} \Delta D$;

Induced effects or "leakages" $\Delta L = B * \Delta X$.

The shock is caused by the modification of an exogenous accounts. The shock causes externalities. It is therefore possible to analyze the impact of shocks on poverty and income distribution on the different social categories. Such work requires that we first convert the SAM into a matrix of average expenditure propensities, and that is substituted for the latter the marginal propensities of expenditure.

Linking multipliers and poverty index

Using the Social Accounting Matrix (SAM), it is possible to analyze the spillover effects of a change in one element of the exogenous accounts on income distribution and poverty. Such an analysis is made from the relationship that is established between the multipliers. Our approach is based on that used by Cisse et al. (2001), originally developed by Thorbecke et al. (1996). To do this, under matrix, linking household accounts and production accounts, is extracted from the SAM. A_{mp} under this matrix is also the product of two matrices, namely the effects of interdependence R_{mm} and the distributional effects D_{mp}. The following relationship can be established:

$a_{ij} = r_{ij} * d_{ij}$

with

a_{ij} an element of A_{mp}

r_{ij} an element of R_{mm}

d_{ij} an element of D_{mp}

The impact of another policy on poverty is then apprehended from the following relationship.

$dP_{\alpha ij} / P_{\alpha ij} = a_{ij} \eta_{\alpha i} (dX_j / Y_i) = r_{ij} d_{ij} \eta_{ij} (dX_j / Y_i)$ (1)

Où

dX_j = variation of endogenous accounts;

η_{ij} = Elasticity of FGT poverty indices in relation to average income;

Y_i = Average income of the socio-economic group i;

P_α = FGT poverty class index.

S.A.M year base

The study will be conducted over the year 2005 as reference for the construction of the SAM. The choice of this year is due to the fact that 2005 is considered in national accounts as a year after the normal drop in economic growth observed in 2002. This date also corresponds to the first phase of the implementation of the strategy for reducing poverty and developing an action plan for accelerated growth strategy in Senegal.

Figure 1: Shocks transmission mechanisms between different accounts of the SAM

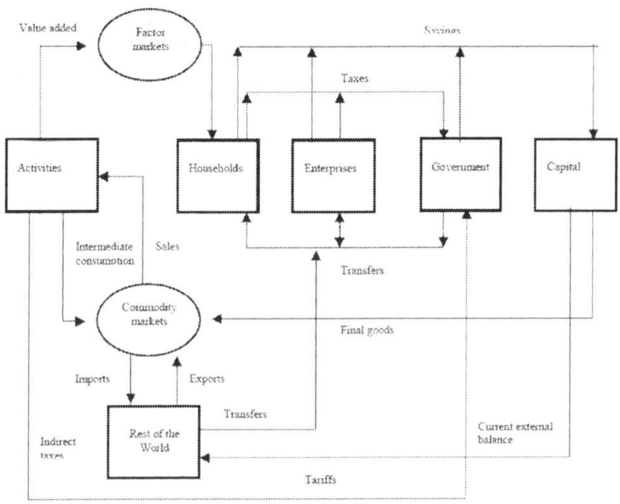

To determine the impact of ICT investment on employment, we use two approaches. The first is a sectoral approach based on the microeconomic behaviour of households. And the second is based on a meso economic economy.

(1). We seek the elasticity of employment relative to wages by each sector of the economy. This relation takes the following form:

Whatever the sector i, we have: log(Popact_i) = α + β*log(Remsal_i) with Popact and Remsal representing respectively the labor force and compensation of employees in the sector i. The econometric estimation of this equation over the period 1980-2005 as log information on the dynamics of each sector in employment.

(2). We then use data from household surveys to determine the proportion of average wages paid in each sector relative to the average wage of the population. Estimates at this level are performed on the 1994 Senegalese national household survey. Let M be the average wage of the population and m_i the average wage of the population in sector i, we calculate $\gamma_i = m_i / M$ where $\sum \gamma_i \neq 1$.

This ratio gives a breakdown of the overall impact achieved by using the SAM multiplier model at each branch i.

As ϖ is the impact of increased investment on ICT for a given economic agent on the labour remuneration. The number of new jobs created all things being equal in the industry i is given by the formula:

$$\varphi_\iota = \gamma_i * \beta_i * \varpi$$

φ_i is the induced effect of the investment decision taking by economic agent i on the creation of employment in the sector i,

The other parameters γi, βi, ϖ, represent respectively, the average weight wage of the branch j compared to the overall average wage, the elasticity of wage earnings in the branch j relative to the workforce the same branch and finally, the impact of increased investment on ICT of an economic agents (households in the formal sector, informal households, businesses, government and rest of world) divide by the labour remuneration.

5. Results

Simulation of a 10% increase of ICT investment

The analysis of the impact of ICT investment on the fundamentals of the economy shows that the increase of 10% of ICT investment leads directly to formal household increased from 0.47% of demand in the goods market, with 0.06% of agricultural products, 0.22% for industrial products, 0.15% for services and 0.04% for non-market services. This additional demand has a positive effect on the supply with an amplitude of 0.29% on the activities offered with a breakdown in agriculture (0.03%) at the industrial level (0.10%), the activities services (+0.12%) and the activities of non-market services (+0.04). Given the openness of the economy on the rest of the world and the structure of production and consumption extrovert economy, the demand generated at local benefits to the rest of the world knows that a net improvement of income 0.08%. The additional supply has created an impact on the value added at factor cost by 0.14%, respectively, with a contribution of 0.09% and 0.05% of production factors labour and capital.

This situation results in an increase in household income of 0.41%, respectively (0.19% for households and 0.22% for formal and informal households). The extra household income is used to pay the taxes thus constituting an increase in income of the Government, and to save. In the same wake, a part of capital income is used to increase the income levels of business, 0.05% and pay taxes to the state and save. These households and businesses pay extra taxes which, together with indirect taxes, an increase of 0.06% of state revenue.

Households in the informal sector

For informal households, their additional investment of 10% in ICT is reflected in the goods market by increasing demand for 14.03%, with 1.70% for agricultural products, 6.69% for industrial products, 4.42% for services and 1.22% for non-market services. Reacting to the rising demand for goods, supply of activity increased from 1.03% for agricultural products, 2.98% for industrial products, 3.50% for services and 1.22% for non-market services for a total of 8.73% which influences the demand for intermediate goods. According the strong dependence of the Senegalese economy vis-à-vis the rest of the world in consumer and production goods, the increase of domestic demand benefit directly the suppliers countries by boosting their net income (+2.39%).Requiring many more factors, the additional supply causes an increase of 4.12% of the value added (respectively with a rising of +2.55% of labor input and +1.57% in capital.

The income of production factors associated with those transferred from the outside, the government and companies improve the households of 12.06% with 5.42% for households and 6.64% for formal and informal households. The additional income of households and firms (1.41%) depending on transfer payments from the rest of the world and the government are used to save to finance, with the savings of the rest of the world, a new household investment and, combined with indirect taxes to increase from 1.68% the income of the State.

Business

Simulating the impact of business investment showed an increase of 10% of ICT investment leads the market for goods increased demand for industrial products of 3.29%, while applications of products increased 0.84% for agricultural products, 2.19% for services and 0.61% for non-market services. This trend of 6.93% of total demand causes a higher offer that influence the supply of activity of 4.32% from agricultural products (0.51%), industrial products (1.47%), services (1.73%) and non-market services (0.61%). This increased activity affects the remuneration of factors of production which has increased from 2.03% of remuneration consisting of labour (1.26%) and capital (0.77%). This application also causes an increase in higher imports than exports, resulting in a gain of outside income estimated at 1.18% of which will be transferred to households. The combined effect of this transfer and remuneration of factors of production is to increase household incomes and formal 2.69% informal household income of 3.23%, a total of 5.92%. Part of the additional household income and surplus income of companies (0.76%) will be spared and influence business investment. Such income and indirect taxes will cause an increase of 0.83% of government income.

The rest of the world

The increase of 10% of foreign direct investment is spread on the goods market in an increase of 8.86% of the application consists of 1.07% of agricultural products, 4.22% of industrial products, 2.80% of services revenues and 0.77% for non-market services. This in turn boosts demand supply which resulted in an estimated increase to 5.56% of the activities offered, including 0.65% and 1.91% respectively for agricultural and industrial activities and 2.22% and 0.78% for service activities and non-market. It also weighs favourably on income from the rest of the world rose 1.59%.

The increased activity leads to an increase offered value-added inputs of 2.62%, up from 1.62% of wages, and 1.00% for capital. The positive trend in the remuneration of production factors and the external and public transfers increase household income 7.51% 3.42% which represents the formal household income, 4.09% of household income informal. The combination of all these activities results in an increase of 0.89% of corporate earnings. Households and companies are charging a portion of their income they pay in taxes direct and indirect taxes from the activities offered, where a change of 1.07% of state revenues.

Government

The 10% increase in public investment is reflected by an increase in demand for goods of 8.38%, with increases of 1.00% of agricultural products, industrial products 3.96% 2.66% services and 0.76% of non-market services. Demand determines the supply of activity and the additional income earned by the rest of the world estimated at (1.42%).

Indeed, the range of activities has increased from 5.24% with an increase of 0.61% of the activities offered at the agricultural sector, 1.76% of supply of industrial, 2.10% of offer of services and activities offered 0.77% of non-market services activities. The increased supply of activity has a positive effect on the value added of production factors which improved from 2.46% consecutively after work (1.53%) and capital (0.93 %). The remuneration of production factors is transmitted with transfers from outside the government and the income of households increased by 7.08%. The income of households and businesses that depend on transfers from the state will be used to pay taxes to the government, resulting in an increase of 1.08%. The derivation of the economic agents on ICTs investment impact on employment is shown in the table above.

Derivation of the impact of ICT investment on employment

Table 6: Derivation of ICT Investment Impact on Employment

Sector	10% Formal Household	10% Informal Household	10% Firms	10% Government	10% RDM
Agriculture, Livestock and Forestry	32	900	445	540	572
Fishing	15	432	214	259	275
Extraction	22	610	302	366	388
Manufacture of food products, beverages and tobacco	93	2625	1297	1575	1668
Manufacture of fabrics, leather and textiles goods	13	370	183	222	235
Manufacture and repair of wooden structures	13	374	185	224	237
Manufacture of paper, printing and publishing	0	-7	-3	-4	-4
Chemical Manufacturing	0	0	0	0	0
Manufacture of building materials	5	141	70	84	89
Manufacture and repair of machinery and fabricated metals	-1	-15	-7	-9	-9

Manufacture of miscellaneous items	2	67	33	40	43
Water, Electricity and Gas	47	1343	664	806	853
Buildings and Construction	18	521	257	312	331
Trade	183	5182	2560	3109	3292
Restaurants and hotels	8	217	107	130	138
Transports and communications	5	150	74	90	95
Banks and financial institution	6	169	83	101	107
Others market services1	-6	-160	-79	-96	-102
Public administration	722	20445	10102	12267	12989
Total new jobs	1178	33364	16486	20018	21196

Source: Author estimations

Overall, the magnitude of the impact is greater when the investment decision from informal household, the rest of the world, the state and enterprises. The impact on employment is lower when the investment decision from households formal.

In total, an increase of 10% of investments ICT, respectively, of households formal, informal households, businesses, state and the rest of the world, can generate, 1178, 33,364, 16,486, 20,018 and 21,196 jobs. In the following, we analyze the impact of the ICT investment of 10% on the job considering the most important answers (informal households and rest of world).

Households in the informal sector

Simulating the impact of ICT investment in informal households resulted in an increase in the supply of services especially trade and branch administration. Reacting to this new demand, it follows an increase in labour demand. The new jobs will stabilize at 20,445 jobs for branch administration and 5182 for the trade branch. This gives new results in increased household incomes thereby stimulating consumer demand causing the reaction of industries producing goods of first necessity. Thus, we see 2625, 1343 and 900 new jobs created respectively for the manufacturing industries of food, beverages and tobacco, water, electricity and gas and agriculture., Livestock and forest.

Note also the perverse effects of this investment on certain sectors such as Other Services, Merchants and Manufacturing and repair of machinery and metal work). Indeed, the acquisition of new equipment and increasing productivity of physical capital will result in a loss of activities of labour associated with declining demand for services and the substitution effect.

Other services branch, Merchants are provided with the least loss in the number of employees of 160 and 15 respectively. Overall this investment has enabled the creation of 33,364 jobs in the national economy.

The rest of the world

Rising foreign direct investment, creating an increased demand for ICT has been able to stimulate the supply of services especially in the sector trade and public administration to respond to new demands of the RDM. The same labour cost required to meet the new demand has allowed the creation of 12,989 jobs for 3292 branch administration and for the trade branch. The new jobs created have, in turn, improved household incomes thereby stimulating consumer demand, which led other industries related to production of first necessity to react to the new demand by increasing their output of leading cost to the stimulation of demand for labour. Thus, there is 1.668, 853 and 572 new jobs created respectively for the manufacturing industries of food, beverages and tobacco, water, electricity and gas and agriculture, livestock and forest.

On the other side of this investment has had a negative impact on certain sectors such as Other Merchant Services, and Manufacturing and repair of machinery and fabricated metals. Indeed, the acquisition of new equipment will result in a loss of repair activity and causing job loss from these branches 102 and 9 respectively for dealers and other services Manufacture and repair of machines and metal products.

Overall this investment has enabled the creation of 21,196 jobs in the national economy all branches.

From the standpoint of the number of jobs created in the national economy, a policy for more than 10% of ICT investments of households working in the informal sector would be most effective. But in terms of population distribution between the different branches, there seems to be an inequality in the distribution of jobs created. The changing population in the primary sector constitutes more than half of the total population with poverty rate of 63.86%, 44.27%, 35.33% respectively for the branches Agriculture, Livestock, Forestry, Fishing and extraction does receives only 5.82% of jobs created or a number of 194.

The industry and public administration among the most favoured with a poverty rate of 15.30% benefits alone more than half the jobs created or 61.27%.

Impact of ICT investments on poverty

The effect of a 10% increase in ICT investments, respectively, formal households, those informal, businesses, state and around the world, shows that in aggregate the observed impact is greatest in households not poor than poor households. However, investment decisions of economic agents mentioned above have a positive impact in reducing poverty. Thus, we note respectively for households:

- Improved revenue of 0.01% for the poor and the non poor, for investment for households TICS formal 10%
- Increase in the incomes of the poor of 0, 30% and 0.34% for non-poor, investment in ICT household moving in the informal sector
- Increased income of the poor by 0.15% and 0.17% for non-poor, for a business investment in ICT by 10%
- Increased income of the poor by 0.18% and 0.18% for non-poor for a state investment in ICT by 10%
- Improving the incomes of poor by 0.19% and 0.21% for the non-poor, with an investment from the world of 10%.#

Depending on the type of employment, the table above shows that the impact of investment is much more pro-poor at the household level that works for the government and NGOs. The impacts observed are respectively:

- For simulation on a 10% increase in household investment ICT formal government workers noted an increase in the non-poor income of 0.01% and 0.02% for non-poor. For households who work for NGOs, there is an almost equal the impact of investment (0.01%).
- For simulation on a 10% increase in ICT capital informal households, the income of government workers increased by 0.31% respectively for non-poor and 0.43% for poor households working for NGOs have also increased similar but less pronounced. Indeed, the income of workers of NGOs is improving, respectively, 0.24% for the non-poor and 0.33% for the poor.
- For simulations on a 10% increase in ICT investments, respectively, businesses, state and the Rest of the world, the trend observed above is preserved. In descending order, the amplitudes of the impacts of these investments on the income of households who work for the government and NGOs, are higher for simulations on: The Rest of the world, the State and enterprises. The simulations show that the impact of in-

vestments ICT has more to households not poor workers working in the public sector Para, Private Company, Embassies and International Organizations, individuals / private households.

6. Conclusion

The world is currently experiencing unprecedented transformation in its history. These changes are mainly caused by the double impact of information technology and communication. Faced with profound changes that characterize the ICT sector of telecommunications is to upset the production processes and patterns in all countries of the world, and particularly in Senegal. The analysis of the impact of ICT investment on the fundamentals of the economy has shown dynamism and scope for the informal sector compared to other sectors. The informal households, following an increase of 10% of ICT investments generate much more revenue. Strength of this finding, it will be much wiser to promote, encourage investment at the household level in the informal sector in line with the strategies for poverty reduction and accelerated growth strategy.

As in most SSA countries, the informal sector plays a very important role in the economy of Senegal. Indeed, a significant fraction of the population lives in what are called "small jobs". This sector generates significant new jobs often precarious in most developing countries. However, we must recognize that their activities are beyond any kind of taxation, control or even just counting in general. Unable to fight head on with a sector as dynamic and ubiquitous, the government should try to encourage the informal sector to gradually enter the legal fold in order to receive benefits and pensions.

Already, ICTs are a major source of growth and jobs in emerging economies. Investment and promotion on ICT by economic agents generates a lot of positive externalities. From another perspective, the introduction of ICT significantly disrupts the exercise of certain professions. Indeed, new jobs will appear while others will be called to certain demise. Among these may be mentioned that a bookseller highly threatened by the spread on the Net virtual libraries. In turn, trade distribution of music products (records, CDs, etc.) is being reorganized.

References

Acero, L., (1995), 'Conflicting demands of new technology and household work: women's work in Brazilian and Argentinian textiles', in: Mitter, S. and Rowbotham, S. (eds.), *Women Encounter Technology: Changing Patterns of Employment in the Third World*, London and New York: Routledge.

Adelman, I. and Robinson, S. (1978), *Income Distribution Policy in Developing*

Countries: A Case Study of Korea, Stanford, CA: Stanford University Press.

Ayres, R, U. (1991), *Computer Integrated Manufacturing*, vol.1, London: Chapman & Hall.

Bassanini, A. and Scarpetta, S. (2002), 'Growth, Technological Change, and ICT diffusion: Recent Evidence from OECD Countries', in *Oxford Review of Economic Policy*, 18(3): 324–44.

Becchetti, L. and Adriani, F. (2005), 'Does the Digital Divide Matter? the Role of Information and Communication Technology in Cross Country Level and Growth Estimates', in *Economics of Innovation and New Technology*, 14(6): 435–53.

Brynjolfsson, E. and Hitt, L. (1996), 'Paradox Lost? Firm-Level Evidence on the Returns to Information Systems Spending', in *Management Science*, 42 (4): 541-558.

Dedrick, J., Gurbaxani, V. and Kraemer, K. L. (2003), 'Information Technology and Economic Performance: a Critical Review of the Empirical Evidence', in *ACM Computing Services*, 35(1): 1–28.

Defourny, J. and Thorbecke, E. (1984), 'Structural Path Analysis and Multiplier Decomposition', *Economic Journal*, 94 (March): 111-136.

Doms, M., Dunne, T. and Troske, K. R. (1997), 'Workers, wages and technology', in *Quarterly Journal of Economics*, CXII (February): 253-90.

Dutta, D. and Otsuka, K. (2004), 'Knowledge Spillover from Information and Communication Technology: A Comparative Study of Australia, Japan, South Korea and Taiwan', *Working Paper ECON2004-8*, University of Sydney, School of Economics and Political Science.

EC (2005), *i2010 – A European Information Society for Growth and Employment*, Brussels, Belgium: Commission of European Communities.

Freeman, C. and Soete, L. (1985), *Information Technology and Employment: An Assessment*, Brighton, UK: Science Policy Research Unit, Sussex University.

Fuss, M. and Waverman, L. (2005), 'The Networked Computer: the Contribution of Computing & Telecommunications to Economic Growth Productivity OR Why is there no New Economy in Old Europe: A Production Function Approach', *Working Paper No. 1*, London Business School, Digital Transformations.

James, J. (1994), 'Microelectronics and the the third World', in Cooper, C. (ed.), *Technology and Innovations in the International Economy*, Tokyo: United Nations University.

Jorgenson, D. W. and Stiroh, K. J. (2000), 'Raising the Speed Limit: U.S. Economic Growth in the Information Age', in *Brookings Papers on Economic Activity*, 1: 125–211.

Jorgenson, D. W., Ho, M. S. and Stiroh, K. J. (2003), 'Lessons from the U.S. Growth

Resurgence', in *Journal of Policy Modelling*, 25(5): 453–70.

Kuwahara, Y. (1984), 'Creating New Jobs in High-Technology Industries', *Paper presented at OECD Inter-Governmental Conference on Employment and Growth in the context of Structural Change*, Paris, OECD.

Lawrence, S. (1984), 'Resource constrained project scheduling: an experimental investigation of heuristic scheduling techniques (supplement)', *Technical Report*, Graduate School of Industrial Administration, Carnegie-Mellon University, Pittsburgh, Pennsylvania.

Lee, T. S. Y. and Guo, X. (2004), 'Information and Communications Technology (ICT) and Spillover: a Panel Analysis', *paper presented at Econometric Society Far Eastern Meeting 2004*, 30 June-2 July, Seoul.

Lehr, B. and Lichtenberg, F. (1999), 'Information Technology and Its Impact on Productivity: Firm-Level Evidence from Government and Private Data Sources, 1977-1993', in *Canadian Journal of Economics*; 32(2): 335-62.

Leontief, W. and Duchin, F. (1983), *The Impact of Automation on Employment, 1963–2000*, New York: Institute for Economic Analysis, New York University.

McGuckin, R. H., Spiegelman, M. and van Ark, B. (2004), 'The U.S. Advantage in Retail and Wholesale Trade Performance: How Can Europe Catch up?', *Paper presented at EPKE Final Conference on Information Technology, Productivity and Growth*, 28-29 October, London.

Nordhaus, W. (2005), 'The Sources of the Productivity Rebound an the Manufacturing Employment Puzzle', *NBER Working Paper No. 11354*, National Bureau of Economic Research, Cambridge, MA.

O'Mahony, M. and van Ark, B. (2003), *EU Productivity and Competitiveness: an Industry Perspective - Can Europe Resume the Catching-up Process?*Luxembourg: Office for Official Publications of the European Communities.

O'Mahony, M. and Vecchi, M. (2005), 'Quantifying the Impact of ICT Capital on Output Growth: A Heterogeneous Dynamic Panel Approach', in *Economica*, 72(288): 615–33.

Oliner, S. D. and Sichel, D. E. (1994), 'Computers and Output Growth Revisited. How Big is the Puzzle?', in *Brookings Papers on Economic Activity*, 25(2): 273-334.

Oliner, S. D. and Sichel, D. E. (2000), 'The Resurgence of Growth in the Late 1990s: Is Information Technology the Story?', *Journal of Economic Perspectives*, 14(4): 3-22.

Oulton, N. and Srinivasan, S. (2005), 'Productivity Growth and the Role of ICT in the United Kingdom: an Industry View, 1970-2000', *CEP Discussion Paper No. 681.24*, (CEP Discussion Papers CEPDP0681). Centre for Economic Performance (CEP), The London School of Economics and Political Science (LSE): London, UK.

Peitchinis, S. (1984), *Computer Technology and Employment: Retrospect and Prospect*, London: St. Martin's Press.

Pilat, D. (2004), 'The Economic Impacts of ICT - What Have We Learned Thus Far?', *Paper presented at the 4th ZEW Conference* on the Economics of Information and Communication Technologies, Mannheim, July 2-3.

Pyatt, G. and Round, J. (eds.) (1985), *Social Accounting Matrices: A Basis for Planning, Washington*, DC: The World Bank.

Pyatt, G. and Roe, J. (1977), *Social Accounting for Development Planning*, Cambridge: Cambridge University Press.

Pyatt, G. and Round, J. (1979), 'Accounting and Fixed Price Multipliers in a Social Accounting Matrix Framework', in *Economic Journal*, 89(356): 850-873.

Pyatt, G. and Round, J. (eds.) (1985), *Social Accounting Matrices: A Basis for Planning*, Washington, D.C.: The World Bank.

Rada, J. F. (1982), *The impact of microelectronics and information technology: Case studies in Latin America*, Paris: UNESCO Publication, Vendome, Imprimerie Presses Universitaires de France.

Rahim, S. A. and Pennings, A. J. (1987), *Computerisation and Development in South East Asia*, Singapore: Asian Mass Communications Research and Information Centre.

Roach, S. S. (1989), 'America's White-Collar Productivity Dilemma', in *Manufacturing Engineering*, August, p. 104.

Robinson, S., and Roland-Holst, D.W. (1988), 'Macroeconomic Structure and Computable General Equilibrium Models', in *Journal of Policy Modelling*, 10: 353-377.

Roland-Holst, D. and Sancho, F. (1992), 'Relative Income Determination in the United States: A Social Accounting Perspective', in *Review of Income and Wealth*, 38 (3): 311-327.

Round, J. I. (1985), 'Decomposing Multipliers for Economic Systems Involving Regional and World Trade', in *Economic Journal*, 95: 383-399.

St. Hillaire, F. and Whalley, J. (1983). 'A Microconsistent Equilibrium Data Set for Canada for Use in Tax Policy Analysis', in *Review of Income and Wealth*, 29 (June): 175-204.

St. Hillaire, F. and Whalley, J. (1987), 'A Microconsistent Data Set for Canada for Use in Regional General Equilibrium Policy Analysis', in *Review of Income and Wealth,* 33(3): 327-343.

Sichel, A. (1997), *The Computer Revolution: An Economic Perspective*, Washington, D.C.: Brookings Institution Press.

Sim, C. N. C., and Young, C. (1995), 'Information Technology, Gender and Employment: A Case Study of the Telecommunications Industry in Malaysia', in Mitter, S. and Rowbotham, S. (eds.), *Women Encounter Technology*, London and New York: The United Nations University Press.

Stiroh, K. J. (2002a), 'Are ICT Spillovers Driving the New Economy?', in *Review of Income and Wealth*, 48(1): 33–57.

Stiroh, K. J. (2002b), 'Information Technology and the U.S. Productivity Revival: What Do the Industry Data Say?', in *American Economic Review*, 92(5): 1559–76.

Stone, R. (1981), *Aspects of Economic and Social Modelling*, Geneva: Librairie Droz.

Talero, E. and Gaudette, P. (1995), 'Harnessing information for development: A proposal for World Bank group strategy', *Discussion Paper No.313*, The World Bank, Washington D.C.

Timmer, M., Ypma, G. and van Ark, B. (2003), 'IT in the European Union: Driving Productivity Divergence?', *Research Memorandum GD-67*, Groningen Growth andDevelopment Centre, Groningen, October.

Chapter 12
Impact of Cellular Phones on Technical and Allocative Efficiencies of Commercial Taxi and Fishing Boat Firms in Ghana

Vijay K. Bhasin, Simon K. Harvey & Eme Umoeka Fiawoyife

1. Introduction

In 1998 OECD member countries agreed to define the Information and Communication Technology (ICT) sector as a combination of manufacturing and services industries that capture, transmit and display data and information electronically. The ICT breaks the traditional dichotomy between manufacturing and services making it possible for activities producing and distributing ICT products to pervade all sectors of the economy (OECD, 2002). The definition, thus, paves way for understanding the multi-dimensionality of the ICT and its applicability across various sectors. According to Wangwe (2007) three broad sub-systems make up the ICT system, each of which comprise further subsystems: technology and infrastructure (comprising communication and data processing); institutions (comprising legal and economic institutions) and social structures and processes (these influence ICT usage and the inter-personal links that facilitate the transfer of information). The ICT systems are pervading virtually all forms of human endeavour: work, education, leisure, communication, production, distribution and marketing as well as the content of information networks. ICT plays an important role in determining levels of development and the quality and accessibility of services in our modern societies. The role of ICT in economic development is expected to be manifested in productivity increase, enhancing the quality of life, reducing prices, creating new economic activities and new employment activities as well as generating wealth (Pohjola, 2001). The future of ICT in the African continent depends upon its ability to develop their skills to cope with the changing ICT environment.

It is generally argued that ICT have the potential to reduce poverty and improve livelihoods skills for increasing productivity (Kenny, 2002). In an information age, information asymmetries are one of the major causes for high transaction costs, uncertainty and therefore market failure. A reduction of information asymmetry will also create new opportunities and therefore enhance the efficiency of resource allocation (Akerlof, 1970). The mobile technology brings communications to whole communities that previously had little or no access to fixed line telephones. Mobile telephone services have proved to be easier to access and more flexible to deploy than fixed line communications. Furthermore, mobile coverage delivers a basic infrastructure of communication to communities that road, rail and other communications infrastructure cannot reach as easily. Mobile phones have economic importance for many users in developing countries, as they are enablers of business, in particular for micro-entrepreneurs. Artisans advertise themselves by giving a mobile phone number, taxi drivers are contacted by phone, and retailers do pre-shopping over the phone for supplies. A large majority of small businesses use mobile phones, many of them had no form of telephone access before the acquisition of a mobile phone. Banks are offering cash transfer services through mobile phones in order to avoid queues. The use of phones has resulted in increased turnover and greater efficiency in business.

By the end of December 2008, there were five (5) cellular network operators. SCANCOM (Ghana) Limited (SPACEFON/AREEBA/MTN) is the leading mobile phone provider with a national outlook and about 6.428 million subscribers as against 2.888 million subscribers of MILLICOM (Ghana) Limited (MOBITEL/TIGO), which has about 60% national coverage. GHANA TELECOM (ONE TOUCH/VODAFONE) also has a national outlook, but experiences customer satisfaction difficulties at the moment. By September 2008, GHANA TELECOM had 1.671 million subscribers. By December 2007, KASAPA Limited (formerly CELLTEL) had 0.289 million subscribers. By December 2008, Zain had 0.270 million subscribers. The country had 10.24 million subscribers in total or a 43.81 penetration rate as by December 2008. According to World Bank (2009), there were 502 fixed line and mobile subscribers per 1,000 people. There were 18.1 internet users per 1000 people in 2005 and it increased to 43 per 100 people in 2008.

In Ghana, the taxi industry operates on a small scale. The commercial taxi firms can use the cellular phones to direct the drivers to pick up passengers from different locations, know the prices of spare-parts and taxis, contact the fitters for the repair of taxis, other taxi owners and police, and for the sale and purchase of taxis. Moreover, drivers can communicate with the owners of taxis in case of breakdown, one driver can also communicate with other drivers and taxi drivers can communicate with passengers. Thus the use of cellular phones by the commercial taxi firms as well as taxi drivers can increase their productivity and reduce the transaction costs.

The fishing-boat firms can use the cellular phones to sell their fish, know the location where to sell their fish catch, know the price of fish at various locations, communicate with boats when they are on the high sea and direct them where to go and catch fish, divert the sale of fish to location where they can fetch higher price, and to sell fish to other boats at high sea. The fishing boats can also communicate with each other on the high sea and move to a location where they can get better fish catch. Again, the use of cellular phones by the fishing-boat firms as well as fishing boats can increase their productivity and reduce the transaction costs.

It is important for the policy makers to know the productivity of a factor of production, know the levels of technical and allocative efficiencies, and their significant determinants so that the policy makers can know on which factor or determinant to concentrate in the transport and fishing industry in Ghana.

2. Literature Review

Evidence about the contribution of ICT investment to productivity has been very controversial. In developed countries and especially among the G-7 countries, ICT investment has a large impact on productivity growth in the United States; while in Japan, the United Kingdom and France, for example, labour productivity did not increase despite a high level of investment in ICT (IMF, 2001).

The relationship between ICT and productivity has long been debated over the past three decades. In the 1980s and in the early 1990s, empirical research generally did not find relevant productivity improvements associated with ICT investments (Strassmann, 1990; Lovemann, 1988; Roach,

1989). This research showed that there was no statistically significant, or even measurable, association between ICT investments and productivity at any level of analysis chosen. Robert Solow commented this puzzle by saying that "we find computer everywhere, except in the productivity statistics." Various rationales have been suggested to explain this paradox. First, simple bivariate correlations between aggregate productivity and aggregate ICT capital stock do not take into account the impact of all controls which also affect aggregate productivity and are therefore likely to measure spurious effects (Lehr-Licthemberg, 1999). Milgrom-Roberts (1988) show for instance that computer aided design (CAD) reduces costs of adjusting products to changing consumer tastes. This shifts productive organization from the Fordist approach to a more flexible network approach which emphasizes economies of scope more than economies of scale. Since variety is hard to detect than quantity it is difficult to find trace of these changes in data. Third, productivity gains from ICT investment materialize only after time and depend significantly on network externalities and on changes in the complementary infrastructure (David, 1990). Fourth, output measurement errors may affect estimates of the impact of ICT investment on output as quality improvements in products and in services are not fully reflected in sales. Fifth, ICT accounts for a relatively small share of capital output so that its increase has only small effects on aggregate output.

More recently, as new data were made available and new methodologies were applied, empirical investigations have found evidence that ICT is associated with improvements in productivity, in intermediate measures and in economic growth (Oliner and Sichel, 1994; Lehr-Licthemberg, 1999; Sichel, 1997; Brynjolfsson and Hitt, 1996). An interpretation consistent with older and more recent results is that micro level data, by allowing the use of more controls, succeed better than aggregate measures in isolating ICT effects on productivity from changes in other conditioning variables. A first direction for improving the last vintage of empirical approaches is disaggregation of the impact of ICT into the contributions of components with different effects and characteristics (such as software, hardware and telecommunications). A second direction is going beyond the adoption of two-stage estimation procedures (production function and then separately an estimation in which the dependent variable is the residual from the first equation) which inconsistently assume the independence of the inefficiency effects in the two estimation stages and are likely to be inefficient.

Following this approach, Paganetto, Becchetti and Bedoya (2000) analyze the determinants of ICT investment and the impact of information technology on productivity and efficiency on a representative sample of small and medium sized Italian firms. In order to test different theoretical predictions from the ICT literature they evaluate the impact of investment in software, hardware and telecommunications of these firms on a series of intermediate variables and on productivity. Among intermediate variables they consider the demand for skilled workers, the introduction of new products and processes and the rate of capacity utilization. Among productivity measures they include total factor productivity, the productivity of labor, and the distance from the "best practice" by using a stochastic frontier approach. Their results show that the positive effect of ICT investment on firm efficiency can be more clearly detected at firm level data by decomposing it into software and telecommunications investment. They find that software investment increases the demand for skilled workers, average labour productivity and proximity to the optimal production frontier. On the other hand, telecommunications investment positively affects the creation of new products and processes but negatively affects average labour productivity. The authors interpret these results by arguing that ICT investment modifies the trade-off between scale and scope economies. While software investment increases the scale of firm operations, telecommunications investment creates a "flexibility option" easing the switch from a Fordist to a flexible network productive model in which products and processes are more frequently adapted to satisfy consumers' taste for variety.

Atzeni and Carboni (2001) employing a large data set of manufacturing firms in this study and try to assess the impact of information and communication technology (ICT) on productivity growth in Italy and to investigate the differences in ICT adoption between North and South of the country. Not all firms invest in ICT and skills, or better to say, not at the same rate. This situation is found to be at the base of the broad productivity variance across the sample. In Italy performance asymmetry has a strong territorial identity since in the North firms invest more in ICT and ICT complements relatively to the South which appears to be rather backwards. A standard regression methodology is employed to calculate the growth rate of productivity and the impact of ICT adoption on it. They then focus on the different matching between human capital and ICT adoption in the two areas. Their findings support the idea that using ICT has beneficial effects on overall productivity growth. Moreover, the use of ICT can help firms in the South to catch up relatively the northern ones, provided that they employ more skilled workers.

Jorgensen and et. al. (2002) analyzed the sources of U.S. labour productivity growth in the post-1995 period and presented projections for both output and labour productivity growth for the next decade. They found that ICT played a substantial role in the U.S. economy by reviving productivity. Their projections put the rate of trend productivity growth at 2.1 percent per year over the next decade. Daveri (2002) showed that throughout 1992-2001, even if two thirds of the European Union population reached or came too much closer to the same levels of ICT diffusion as the U.S., ICT have so far delivered limited overall productivity gains in Europe. Hempell (2002) found significant productivity effects of ICT on German service sector.

So far as the developing countries are concerned, Muller-Falke (2001) has reported a positive link between ICT capital and productivity by considering small and medium scale enterprises in the manufacturing sector in India. In Africa, a number of studies have been carried out in recent times that focus on industrial and sectoral determinants of ICTs adoption and applications in Africa (Oyelaran-Oyeyinka and Lal, 2004; Moodley, 2002; Oyelaran-Oyeyinka and Adeya, 2004). According to these studies, the adoption of ICTs is likely to promote greater productivity within the enterprises. Chowdhury and Wolf (2002) assessed the uses of information and communication technologies and their impact on the economic performance of small and medium scale enterprises of Kenya, Tanzania and Uganda. Findings suggest that the diffusion of ICT among East African small and medium scale enterprises is both industry and country specific. The model that is based on a Cobb-Douglas specification is modified to take into account ICT impact on labour productivity. The empirical findings suggest that investment in ICT has a negative impact on labour productivity. Nkama (2007) explores the relationship between productivity and investment in ICT in Cameroon at firms' level in 2004. Using cross-section data and applying a Cobb-Douglas function, the study reveals that investment in ICT has no impact on productivity, as the estimated coefficient of ICT investment on productivity is not significant. Also, ICT investment has no impact on labour productivity and labour intensity. These findings differ from results obtained by Chowdhury and Wolf (2002) according to which ICT investment has negative and significant impact on labour productivity in East Africa. In Cameroon labour remains the key factor of value added growth. This seems to be realistic as the country has an important workforce that tends to slow down salaries. Since labour is the abundant factor, it is profitable for firms to increase their production by recruiting more units of labour. If ICT investment contributes to rapid globalization of economies, it does not contribute to productivity growth in Cameroon. There seems to be no empirical literature in Africa that examines the impact of ICT on technical and allocative efficiencies of firms.

3. Methodology

Sampling Procedure and Data Collection

Two Questionnaires were designed to elicit data from the users of cellular phone services, especially fishing-boat and commercial taxi firms located in the Central Region of Ghana. The first challenge that we faced in data collection was to decide from which location to identify the population size. Keeping in view the budget and time, we decided to concentrate on commercial taxi firms located in Cape Coast, Elmina, Moree, Komenda, Mankesim and Fosu, whereas to concentrate on fishing boat firms located in Cape Coast, Elmina and Moree. The commercial taxi drivers were contacted at Cape Coast, Elmina, Moree, Komenda, Mankesim and Fosu to know the addresses of commercial taxi firms. About 320 commercial taxi firms were identified in these localities that use cellular phones to operate their business. Stratified random sampling (lottery method) was used to select 100 commercial taxi firms. Fishing boat operators were contacted at Cape Coast, Elmina, and Moree to know the addresses of fishing boat firms. About 463 fishing boat firms were identified that use cellular phones to operate their business. Again, stratified random sampling (lottery method) was used to select 100 fishing boat firms.

The second challenge that we faced was that the Questionnaires were in English Language and most of the respondents could understand only Fante Language. Enumerators who were very proficient in Fante language were selected and trained for one week so that they could interpret the questionnaires to the selected firms. Before the final questionnaires were administered, questionnaire testing was conducted on five respondents from the selected commercial taxi and fishing boat firms. The testing revealed some weaknesses in the structure of some of the questions in original questionnaire. The questionnaire was therefore modified accordingly and was administered by the enumerators. Answers to the questions listed in the two questionnaires were obtained during April-June, 2009. There was a follow up to confirm some of the responses provided by the respondents.

Frontier Production Function and Technical efficiency estimation

If we are estimating a production function using cross-section data, then we can use the value of output as well as the value of inputs instead of physical quantities of output and inputs because all the firms face the same prices for outputs as well as inputs. Moreover, we should keep in mind that the regression coefficients are not affected by change in origin or scale. Keeping in mind these two properties, we estimate the Cobb–Douglas frontier production function for commercial taxi firms which is specified as:

$$\ln Y_i = \beta_0 + \beta_1 \ln (NICTK_i) + \beta_2 \ln (ICTK_i) + \beta_3 \ln(L_i) + (V_i - U_i) \quad i = 1,2,\ldots,100 \quad (1)$$

where ln represents the natural logarithm; the subscript i represents the ith firm; Yrepresents the output that is approximated by gross sales per month in GH Cedis, NICTKrepresents the non-ict-capital that includes the cost of taxis (fixed capital), expenses on the repair and maintenance of taxis per month and the amount spent on the purchase of petrol/diesel per month (working capital). ICTK represents the ict-capital that includes amount spent on the purchase of mobile phones (fixed capital), amount spent on the purchase of scrach cards/unit transfers per month and amount spent on the repair of mobile phones (working capital). L represents labour that is approximated by expenses on salary of workers per month. The V_i 's are assumed to be independent and identically distributed random errors, having $N(0, \sigma_v^2)$ distribution, independent of the U_is. The U_is are half-normally distributed, which are assumed to be non-negative random variables.

We estimate the Translog frontier production function for fishing boat firms which is specified as:

$$\ln Y_i = \beta_0 + \beta_1 \ln (NICTK_i) + \beta_2 \ln (ICTK_i) + \beta_3 \ln(L_i) + \beta_4 \ln(NICTK_i)^2 + \beta_5 \ln (ICTK_i)^2 + \beta_6 \ln(L_i)^2 + \beta_7 \ln (NICTK_i) * \ln(L_i) + \beta_8 \ln (NICTK_i) * \ln (ICTK_i) + \beta_8 \ln(L_i) * \ln (ICTK_i) + (V_i - U_i) \quad i = 1,2,\ldots,100 \qquad (2)$$

where ln represents the natural logarithm; the subscript i represents the ith firm; Yrepresents the output that is approximated by value of fish catch per month in GH Cedis, NICTKrepresents the non-ict-capital that includes the cost of boats and nets (fixed capital), expenses on the repair of nets per month and the amount spent on the purchase of pre-mix fuel per month (working capital). ICTK represents the ict-capital that includes amount spent on the purchase of mobile phones (fixed capital), amount spent on the purchase of scratch cards/unit transfers per month and amount spent on the repair of mobile phones (working capital). L represents labour that is approximated by expenses on salary of workers per month. The V_i 's are assumed to be independent and identically distributed random errors, having $N(0, \sigma_v^2)$ distribution, independent of the U_is. The U_is are truncated-normally distributed, which are assumed to be non-negative random variables.

The technical efficiency (Te) of the individual firm is defined in terms of the ratio of observed output to the corresponding frontier output, conditional on the levels of input used by the firm. Hence the technical efficiency of firm i is expressed as:

$$Te_i = \ln Y_i / \ln Y^* = \exp(-U_i) \qquad (3)$$

The value of technical efficiency lies between zero and one.

Frontier Cost Function and Allocative efficiency estimation

We estimate the Cobb–Douglas frontier cost functions for commercial taxi and fishing boat firms as suggested by Coelli (1994), which are specified as:

$$\ln (C_i / W_i) = \beta_0 + \beta_1 \ln (Y_i) + \beta_2 \ln (NICTR_i / W_i) + \beta_3 \ln(ICTR_i / W_i) + (V_i + U_i) \quad i = 1,2,\ldots,100 \qquad (4)$$

where C represents the total cost of production, W represents the price of labour, Yrepresents the output, NICTRrepresents the price of non-ict-capital and ICTR represents the price of ict-capital. The V_i's are assumed to be independent and identically distributed random errors, having $N(0, \sigma_v^2)$ distribution, independent of the U_is. The U_is are half-normally distributed, which are assumed to be non-negative random variables.

For commercial taxi firms, the total cost of production includes cost of taxis, expenses on the repair and maintenance of taxis, fuel cost, salaries of workers, costs involved in cellular phones and other costs. The price of labour is the average wage rate. The output is represented by the gross sales received by the commercial taxi firm. The price of non-ict-capital is represented by the price of taxis and the price of ict-capital is represented by the price of mobile phones.

For fishing boat firms, the total cost of production includes cost of boats, cost of nets, expenses on the repair of nets, pre-mix fuel cost, salaries of workers, costs involved in cellular phones and other costs. The price of labour is the average wage rate. The output is represented by the value of fish catch. The price of non-ict-capital is represented by the prices of boats and nets, and the price of ict-capital is represented by the price of mobile phones.

The level of allocative efficiency (Ae) of the individual firm is calculated as the ratio of the predicted minimum cost to the observed cost. This is expressed for each firm as:

$$Ae_i = \exp (U_i) \qquad (5)$$

The value of allocative efficiency lies between one and infinity.

The maximum likelihood estimates for the parameters of the stochastic frontier models (production as well as cost) and the predicted technical and allocative efficiencies are obtained by using the computer programme,

FRONTIER 4.1, in which the variance parameters are expressed in terms of:

$\sigma^2 = (\sigma_U^2+\sigma_V^2)$, and$\gamma = \sigma_U^2/(\sigma_U^2+\sigma_V^2)$ .... (6)

The term γ represents the ratio of the variance of inefficiency's error term to the total variance of the two error terms defined above. The value of γ can range between 0 and 1. The significance of the γ parameter can be used to test whether the stochastic frontier production function and cost function are preferred to the average production function and cost function, respectively. If the null hypothesis, that γ equals zero, is accepted, this would indicate that σ_U^2 is zero and hence that the U_i term should be removed from the model, leaving a specification with parameters that can be consistently estimated using ordinary least squares. The mean levels of technical and allocative efficiencies will represent the technical and allocative efficiencies for the fishing and transport industries.

Technical and allocative efficiencies are assumed to be determined by firm specific variables, and may be expressed as:

$\mu_i = \delta_0 + \Sigma\delta i\ Z, \mu_i = Te_i orAe_i$ .... (7)

where δs are unknown parameters to be estimated and the vector Z includes variables such as age of the owner (age), education of the owner (edu), member or non-member of the association (mem) and ICT investment in cellular phones (ict) that could influence the technical and allocative efficiencies of the firms. Schooling provides knowledge that can enhance the use of cellular phones in the functioning of the taxi and fishing-boat industry. Age gives the experience and it can enhance the utilization of cellular phones in both these industries. If the value of technical efficiency is below one, this shows that some of the firms are operating below the stochastic production frontier. Therefore we want to raise the value of technical efficiency of these firms to the maximum value of one. Since we want to raise the technical efficiency of commercial taxi firms, we expect to observe a negative relationship between the technical efficiency and age of the owner; whereas positive relationship between technical efficiency and the other variables. If the value of allocative efficiency is above one, this shows that some of the firms are operating above the stochastic cost frontier. Therefore, we want to reduce the value of allocative efficiency of these firms to the minimum value of one. Since we want to reduce the allocative efficiency of commercial taxi firms, we expect to observe a positive relationship between the allocative efficiency and age of the owner; whereas negative relationship between allocative efficiency and the other variables. The parameters of these determinants are estimated by using the Ordinary Least Squares (OLS) method. The main hypotheses to be tested are that ICT investment in cellular phones raises the technical efficiency and reduces the allocative efficiency of commercial taxi and fishing boat firms.

4. Empirical Results

Technical and Allocative Efficiencies of Commercial Taxi Firms

The maximum likelihood estimates of preferred frontier production function for commercial taxi firms are presented in Table 1. First of all it is important to note that the Likelihood Ratio test statistic is statistically significant at the 5% level of significance, which implies that the frontier Cobb-Douglas production function (two error terms) fits the data better than an average Cobb-Douglas production function (one error term). The estimated production function exhibits increasing returns to scale. It is evident from the estimated frontier Cobb-Douglas production function for the commercial taxi firms that non-ict-capital and labour are the significant determinants of the value of gross sales. Since the coefficient of the ict-capital is statistically insignificant, this implies that the ICT investment in cellular phones by the commercial taxi firms has no affect on their productivity. This finding is in conformity with the findings of Strassmann (1990); Lovemann (1988); and Roach (1989) for developed countries and Nkama (2007) for Cameroon, a developing country.

The distribution of the estimated technical efficiencies of commercial taxi firms are indicated through histogram in Figure 1. The maximum numbers of commercial taxi firms operate around the mean technical efficiency of 0.63. The spread as well as the values of technical efficiencies indicate that there is a scope for the improvement of technical efficiencies of commercial taxi firms as well as the transport industry. The most important question that arises now is whether the ICT investment in cellular phones can raise the technical efficiency of these commercial taxi firms or not.

The estimates for the determinants of technical efficiency of commercial taxi firms are presented in Table 2. The intercept and the age of the owner of commercial taxi firm are the only significant determinants of technical efficiency of commercial taxi firms. The statistical significance of the intercept shows that there could be some other factors that could raise the technical efficiency of commercial taxi firms. Since age of the owner is a significant determinant of technical efficiency, younger people should be encouraged to start commercial taxi firms. The statistical estimates of the determinants of technical efficiency indicate that the ICT investment in cellular phones is not a significant determinant of the technical efficiency of commercial taxi firms. This finding contradicts the finding of Paganetto, Becchetti and Bedoya (2000).

Table 1: Maximum Likelihood Estimates of Cobb-Douglas Production Function of Commercial Taxi Firms

Y	coefficient	standard-error	t-ratio
ICTK	0.0943	0.1656	0.5696
NICTK	0.3782	0.1419	2.6652*
L	0.7738	0.1526	5.0685*
sigma- squared	0.8596	0.1949	4.4104*
gamma	0.5303	0.1627	3.2583*

mu is restricted to be zero

eta is restricted to be zero

log likelihood function = -113.3886

LR test of the one-sided error = 2.4027**

with number of restrictions = 1

* indicates that the statistic is significant at 1% level of significance

** indicates that the statistic is significant at 5% level of significance

The maximum likelihood estimates of preferred frontier cost function for commercial taxi firms are presented in Table 3. First of all it is important to note that the Likelihood Ratio test statistic is statistically significant at the 1% level of significance, which implies that the frontier Cobb-Douglas cost function (two error terms) fits the data better than an average Cobb-Douglas cost function (one error term). It is evident from the estimated frontier Cobb- Douglas cost function for the commercial taxi firms that the intercept, output, non-ict-capital price/wage ratio and ict-capital price/wage ratio are the significant determinants of the total cost/ wage ratio. Since the coefficient of the ict-capital price/wage ratio is statistically significant, this implies that the ICT investment in cellular phones by the commercial taxi firms has a positive effect on their total cost/wage ratio.

The distribution of the estimated allocative efficiencies of commercial taxi firms are indicated through histogram in Figure 2. The maximum numbers of commercial taxi firms operate around the allocative efficiency value of 1.0. The mean allocative efficiency value is 1.65. The spread as well as the values of technical efficiencies indicate that there is a scope for the reduction of allocative efficiencies of commercial taxi firms as well as the transport industry. The most important question that arises now is whether the ICT investment in cellular phones can reduce the allocative efficiency of these commercial taxi firms or not.

Figure 1: Distribution of Commercial Taxi Owners according to Technical Efficiency

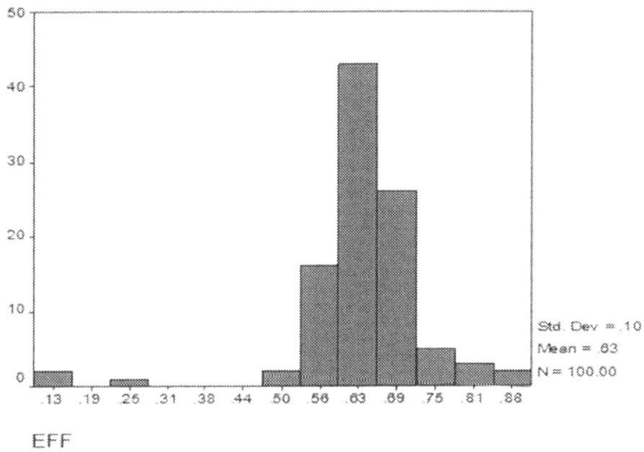

Table 2: Determinants of Technical Efficiency of Commercial Taxi Firms

Te	coefficient	Standard-error	t-ratio	P>t	[95% Conf.	Interval]
age	-.0031187	.0011457	-2.72	0.008	-.0053932	-.0008442
edu	.003408	.0030939	1.10	0.273	-.0027342	.0095503
mem	-.0279054	.0226495	-1.23	0.221	-.0728704	.0170595
ict	-.000217	.0001709	-1.27	0.207	-.0005563	.0001224
_cons	.7607422	.0645933	11.78	0.000	.6325083	.8889761

Table 3: Maximum Likelihood Estimates of Cobb-Douglas Cost Function of Commercial Taxi Firms

C/W	coefficient	standard-error	t-ratio
const	0.3465	0.0293	11.8179*
Y	0.0182	0.0019	9.3906*
NICTK/W	0.9091	0.0034	262.5743*
ICTK/W	0.0635	0.0027	23.4650*
Sigma-squared	0.3126	0.0104	29.9170*
Gamma	0.9999	0.0000	5509733.9*

mu is restricted to be zero

eta is restricted to be zero

log likelihood function = -19.0440

LR test of the one-sided error = 36.3080*

with number of restrictions = 1

* indicates that the statistic is significant at 1% level of significance

Figure 2: Distribution of Commercial Taxi Firms according to Allocative Efficiency

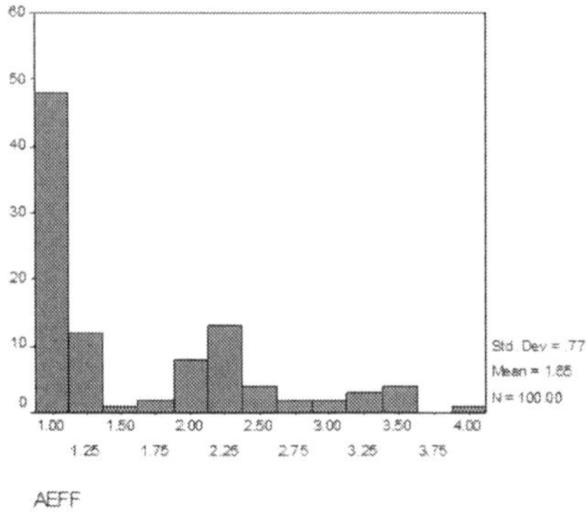

Table 4: Determinants of Allocative Efficiency of Commercial Taxi Firms

Ae	coefficient	Standard- error	t-ratio	P>t	[95% Conf.	Interval]
age	.0224567	.0086272	2.60	0.011	.0053294	.0395839
edu	.0070828	.0232977	0.30	0.762	-.039169	.0533346
mem	.0460682	.170554	0.27	0.788	-.2925242	.3846607
ict	.0026393	.0012872	2.05	0.043	.0000839	.0051946
_cons	.3465339	.4863965	0.71	0.478	-.6190853	1.312153

The estimates for the determinants of allocative efficiency of commercial taxi firms are presented in Table 4. The sign of the estimated coefficient of the age of the owner of commercial taxi firm is as expected "a priori" and it is also statistically significant at less than one percent level of significance. Since the age of the owner is a significant determinant of allocative efficiency, younger people should be encouraged to start commercial taxi firms. The statistical estimates of the determinants of allocative efficiency indicate that the ICT investment in cellular phones is a significant determinant of the allocative efficiency of commercial taxi firms but the sign of the coefficient of this variable contradicts “a priori” expectation. One of the possible reasons for the positive relationship between the allocative efficiency and ICT investment in cellular phones could be that the commercial taxi firms are using very expensive phones to operate their business.

Technical and Allocative Efficiencies of Fishing Boat Firms

Table 5: Maximum Likelihood Estimates of Translog Production Function of Fishing Boat Firms

Y	coefficient	standard-error	t-ratio
const	0.4115	0.9861	0.4173
NICTK	1.1748	0.6979	1.6833*
L	0.0389	0.6979	0.0558
ICTK	0.0502	0.7977	0.0629
$NICTK^2$	0.0204	0.2646	0.0774
L^2	0.0884	0.0728	1,2139
$ICTK^2$	0.0683	0.4136	0.1652
NICTK*L	-0.1131	0.3007	-0.3763
NICTKl*ICTK	-0.1748	0.4039	-0.4328
L*ICTK	0.1280	0.4523	0.2830
sigma-squared	3.9299	1.4686	2.6759*
gamma	0.9321	0.0200	46.5922*
mu	-3.8280	5.6569	-0.6767*

eta is restricted to be zero

log likelihood function = -114.48877

LR test of the one-sided error = 3.5048**

with number of restrictions = 2

* indicates that the statistic is significant at 1% level of significance

** indicates that the statistic is significant at 5% level of significance

The maximum likelihood estimates of preferred frontier production function for fishing boat firms are presented in Table 5. First of all it is important to note that the Likelihood Ratio test statistic is statistically significant at the 5% level of significance, which implies that the frontier Translog production function (two error terms) fits the data better than the frontier Cobb-Douglas production function (two error terms). It is evident from the estimated frontier Translog production function for the fishing boat firms that only non-ict-capital is the significant determinant of the value of gross sales. Since the coefficient of the ict-capital is statistically insignificant, this implies that the ICT investment in cellular phones by the fishing boat firms has no affect on their productivity. This finding is in conformity with the findings of Strassmann (1990); Lovemann (1988); and Roach (1989) for developed countries and Nkama (2007) for Cameroon, a developing country. However, it should be noted that we have estimated the translog production function, whereas Nkama (2007) estimated the Cobb-Douglas production function.

The distribution of the estimated technical efficiencies of fishing boat firms are indicated through histogram in Figure 3. The maximum numbers of fishing boat firms operate around the technical efficiency of 0.69. The mean technical efficiency of the fishing industry is 0.60 which lower than that of the transport industry (0.63). The spread as well as the values of technical efficiencies indicate that there is a scope for the improvement of technical efficiencies of fishing boat firms as well as the fishing industry. Again, the most important question that arises now is whether the ICT investment in cellular phones can raise the technical efficiency of these fishing boat firms or not.

The estimates for the determinants of technical efficiency of fishing boat firms are presented in Table 6. The intercept and the education of the owner of fishing boat firm are the only significant determinants of technical efficiency of fishing boat firms. The statistical significance of the intercept shows that there could be some other factors that could raise the technical efficiency of fishing boat firms. Since education of the owner is a significant determinant of technical efficiency, educated people should be encouraged to start fishing boat firms. The statistical estimates of the determinants of technical efficiency indicate that the ICT investment in cellular phones is not a significant determinant of the technical efficiency of fishing boat firms. This finding contradicts the finding of Paganetto, Becchetti and Bedoya (2000).

Figure 3: Distribution of Fishing Boat Firms according to Technical Efficiency

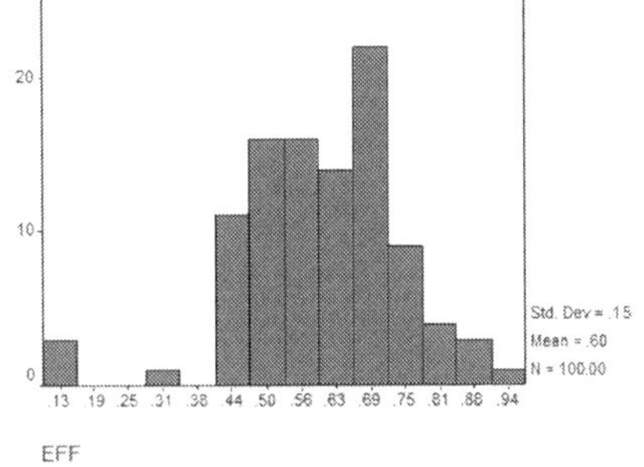

The maximum likelihood estimates of preferred frontier cost function for fishing boat firms are presented in Table 7. First of all it is important to note that the Likelihood Ratio test statistic is statistically significant at the 5% level of significance, which implies that the frontier Cobb-Douglas cost function (two error terms) fits the data better than an average Cobb-Douglas cost function (one error term). It is evident from the estimated frontier Cobb-Douglas cost function for the fishing boat firms that the intercept, output, non-ict-capital price/wage ratio and ict-capital price/wage ratio are the significant determinants of the total cost/ wage ratio. Since the coefficient of the ict-capital price/wage ratio is statistically significant, this implies that the ICT investment in cellular phones by the fishing boat firms has a positive effect on their total cost/wage ratio.

Table 6: Determinants of Technical Efficiency of Fishing Boat Firms

Te	coefficient	Standard- error	t-ratio	P>t	[95% Conf.	Interval]
age	.000525	.0015422	0.34	0.734	-.0025367	.0035867
edu	.0079068	.0030961	2.55	0.012	.0017603	.0140533
mem	.0348802	.0318911	1.09	0.277	-.0284316	.0981919
ict	.0001635	.0002012	0.81	0.418	-.0002358	.0005628
_cons	.4897469	.084967	5.76	0.000	.321066	.6584277

Table 7: Maximum Likelihood Estimates of Cobb-Douglas Cost Function of Fishing Boat Firms

C/W	coefficient	standard-error	t-ratio
const	0.3895	0.1955	1.9919**
Y	0.1288	0.0250	5.1503*
NICTK/W	0.7235	0.0345	20.9138*
ICTK/W	0.1217	0.0324	3.7525*
Sigma-squared	0.1369	0.0346	3.9522*
Gamma	0.8565	0.1052	8.1389*

mu is restricted to be zero

eta is restricted to be zero

log likelihood function = -0.1002

LR test of the one-sided error = 4.4462**

with number of restrictions = 1

* indicates that the statistic is significant at 1% level of significance

**indicates that the statistic is significant at 5% level of significance

Figure 4: Distribution of Fishing Boat Firms according to Allocative Efficiency

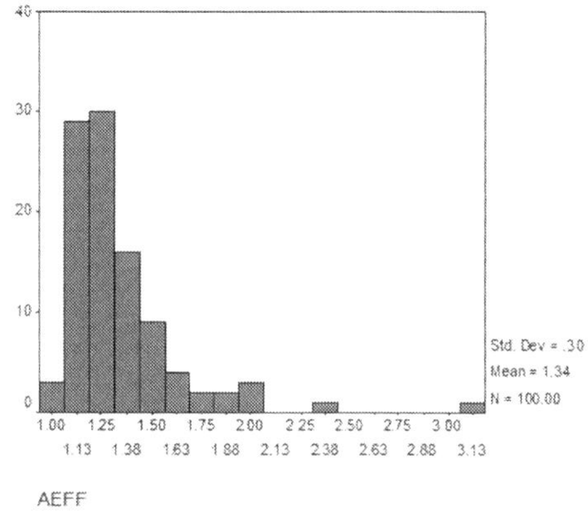

The distribution of the estimated allocative efficiencies of fishing boat firms are indicated through histogram in Figure 4. The maximum numbers of fishing boat firms operate around the allocative efficiency value of 1.25. The mean allocative efficiency value of the fishing industry is 1.34 which is lower than that of transport industry (1.65). The spread as well as the values of technical efficiencies indicate that there is a scope for the reduction of allocative efficiencies of fishing boat firms as well as the fishing industry. The most important question that arises now is whether the ICT investment in cellular phones can reduce the allocative efficiency of these fishing boat firms or not.

The estimates for the determinants of allocative efficiency of fishing boat firms are presented in Table 8. None of the variables appearing in Table 8 have statistically significant coefficients except the intercept. This shows that there may be some other relevant variables that could affect the allocative efficiency of fishing boat firms. The statistical estimates of the determinants of allocative efficiency indicate that the coefficient of the ICT investment in cellular phones has a correct sign but it is statistically insignificant.

Table 8: Determinants of Allocative Efficiency of Fishing Boat Firms

Ae	coefficient	Standard- error	t-ratio	P>t	[95% Conf.	Interval]
age	-.0014481	.0031536	-0.46	0.647	-.0077089	.0048127
edu	.0052991	.006331	0.84	0.405	-.0072696	.0178678
mem	-.063571	.0652126	-0.97	0.332	-.1930343	.0658923
ict	-.0005588	.0004113	-1.36	0.178	-.0013754	.0002578
_cons	1.455521	.1737451	8.38	0.000	1.110593	1.800448

5. Conclusions

The estimates of the stochastic frontier production function show that the ICT investment in cellular phones by the commercial taxi firms does not have any effect on their productivity. However, qualitative analysis shows that the productivity of commercial taxi firms could be raised by using the mobile phones to locate the drivers, drivers communicating with each other as well as with the passengers, and to contact the other taxi-owners. On the other hand, the estimates of stochastic frontier cost function show that the ICT investment in cellular phones by the commercial taxi firms does have some affect on their total cost-wage ratio which implies that the ICT investment is cost effective in transport industry. The estimated values for the technical and allocative efficiencies of commercial taxi firms indicate that there is plenty of scope to raise the technical efficiency and to reduce the allocative efficiency of commercial taxi firms. The significant determinants of technical and allocative efficiencies indicate that this could be achieved by encouraging younger people to start the commercial taxi business.

The estimates of the stochastic frontier production function show that the ICT investment in cellular phones by the fishing-boat firms does not have any effect on their productivity. On the other hand, the estimates of stochastic frontier cost function show that the ICT investment in cellular phones by the fishing-boat firms does have some affect on their total cost-wage ratio which implies that the ICT investment is cost effective in fishing industry. Again, the study shows that there is plenty of scope to raise the technical efficiency and to reduce the allocative efficiency of fishing boat firms. . The mean technical efficiency of the fishing industry is 0.60 which is lower than that of the transport industry (0.63). The mean allocative efficiency value of the fishing industry is 1.34 which is lower than that of transport industry (1.65). Thus fishing industry is less efficient than the transport industry. However, the significant determinants of technical efficiency indicate that the technical efficiency of the fishing-boat firms could be increased by encouraging educated people to start the fishing business.

References

Akerlof, G. (1970), 'The Market for Lemons: Qualitative Uncertainty and the Market Mechanism', in *Quarterly Journal of Economics*, 84(3): 488-500.

Atzeni, G. and Carboni, O. (2001), 'The Economic Effects of Information Technology: Firm Level Evidence from the Italian Case', *Working Paper No. 14*, Centro Ricerche Economiche Nord Sud (CRENoS), University of Cagliari, Italy. Available at: http://www.crenos.unica.it/news/preliminarcnr.html

Brynjolfsson, E. and Hitt, L. (1996), *Paradox Lost? Firm-Level Evidence of High Return to Information Systems Spending*, Available at: http://www.ccs.mit.edu/papers/ccswp 162.

Chowdhury, S. and Wolf, S. (2002), 'Use of ICTs and Economic Performance of Small and Medium Enterprises in East Africa', *Paper Read at WIDER Conference on the New Economy in Development*, Helsinki, Finland.

Coelli, T. J. (1994), 'A guide to FRONTIER version 4.1: A computer program for stochastic frontier production and cost function estimation', *Mimeo*, Department of Econometrics, University of New England, Armidale, Australia.

Daveri, F. (2002), 'The New Economy in Europe (1992-2001)', *Paper Read at Wider Conference on the New Economy in Development*, Helsinki.

David, P. (1990), 'The Dynamo and the Computer: An Historical Perspective on the Modern Productivity Paradox', in *American Economic Review, Papers and Proceedings of the Hundred and Second Annual Meeting of the American Economic Association*, 80(20): 355-361.

Hempell, T. (2002), 'Does Experience Matter? Productivity Effects of ICT in the German Service Sector', *Paper Read at WIDER Conference on the New Economy in Development*, Helsinki.

Jorgenson, D. W., Ho, M. S. and Stiroh K. J. (2002), 'Projecting Productivity Growth: Lessons from the US Growth Resurgence', *Paper Presented at WIDER Conference on the New Economy in Development*, Helsinki.

International Monetary Fund (IMF) (2001), *World Economic Outlook*, Washington, D.C.: IMF.

Kenny, C. (2002), 'Information and Communication Technologies for Direct Poverty Alleviation: Costs and Benefits', in *Development Policy Review*, 20(2): 38-57.

Lehr, B. and Lichtenberg, F. (1999), 'Information Technology and Its Impact on Productivity: Firm-Level Evidence from Government and Private Data Sourced, 1977-1993', in *Canadian Journal of Economics*, 32(2): 335-62.

Loveman, G. (1988), 'An Assessment of the Productivity Impact of Information Technologies', MIT Management in the 1990s, *Working Paper No. 88-05*, July.

Milgrom, P. and Roberts, R. (1988), 'The Economics of Modern Manufacturing: Products, Technology and Organization', *Discussion Paper No.* 136, Centre for Economic Policy Research, Stanford University.

Moodley, S. (2002), 'E-Business in the South African Apparel Sector: A Utopian Vision of Efficiency', in *Developing Economies*, 40(1): 67-100.

Muller-Falke, D. (2001), 'The Use of Telecommunication and Information Technology in Small Business- Evidence from Indian Small Scale Industry', Unpublished Draft.

Nkama, A. (2007), 'Analyzing the Impact of ICT Investments on Productivity

Growth in Developing Countries: Evidence from Cameroon', *Paper Presented at the African Economic Conference*, Addis Ababa, Ethiopia, 15-17 November.

Organisation for Economic Cooperation and Development OECD (2002), *Measuring the Information Economy 2002*, Paris: OECD.

Oliner, S. and Sichel, D. (1994),'Computers and Output Growth Revisited: How Big is the Puzzle?', in *Brookings Papers on Economic Activity*, 2(2): 273-334.

Oyelaran-Oyeyinka, B. and Adeya, C. (2004), 'Internet Access in Africa: Empirical Evidence from Kenya and Nigeria', in *Telematics and Informatics*, 21(1): 67-81.

Oyelaran-Oyeyinka, B. and Lal, K. (2004), 'Sectoral Pattern of E-business Adoption in Developing Countries', *UNU-INTECH, Discussion Paper No.* 7, UNU, Maastricht.

Paganetto, L., Bacchetti, L., Andres, D. and Bedoya, L. (2000), *ICT Investment, Productivity and Efficiency: Evidence at Firm Level Using A Stochastic Frontier Approach*, Available at: http://www.Repec Papers.

Pohjola, M. (2001), *Information Technology, Productivity and Economic Growth: International Evidence and Implications for Economic Development*, WIDER Studies in Development Economics, Oxford: Oxford University Press.

Roach, S. (1989), 'America's White-Collar Productivity Dilemma', in *Manufacturing Engineering*, August, 104: 21-35.

Sichel, A. (1997), *The Computer Revolution: An Economic Perspective*, Washington, D.C.: Brookings Institution Press.

Strassmann, P. (1990), *The Business Value of Computers: An Executive's Guide*, New Canaan, CT: Information Economics Press.

Wangwe, S. (2007), 'Evolution, Status and Impact of ICT on Economic Development and Transformation in Africa: An Overview', *AERC Project on the Impact of ICTs on Economic Development and Transformation*, AERC, Nairobi, Kenya.

World Bank (2009), *Africa Development Indicators*, Washington, D.C.: World Bank

Chapter 13
Impact of ICT Investment, Ownership and Use in the Cassava Value Chain in South Western Nigeria

*Mutawakilu A. Tiamiyu, Abiodun S. Bankole &
Rosemary O. Agbonlahor*

1. Introduction

Meier (2001) has noted that the 'new growth theory' emphasizes the role of ideas in promoting growth through the aggregation of ideas at the micro-level. Such ideas include inventions and their application to productive ends through innovations undertaken by skilled and experienced entrepreneurs. The emergence and rapid global expansion of 'information', 'knowledge' and 'networked' societies is clear confirmation of both the experienced and expected impact of knowledge creation, information sharing, research and development, and innovation on development processes, as well as the role that access to and use of Information and Communication Technologies (ICT) can play in such processes. ICT are the electronic technologies that are used to create, store, transmit, exchange or deliver information. ICT include such technologies as radio, television, mobile phones, satellite systems, computer and network hardware and software, as well as the services associated with these technologies.

ICT are increasingly expected to play key roles in efforts by developing countries to reduce their poverty rates because of the opportunities that they may provide for increasing the incomes of the poor and other disadvantaged groups and enhancing overall national social and economic growth. This expectation is often based on the observed high correlation between the development and use of ICT infrastructure and economic growth and prosperity (measured by GDP, profitability and incomes at the person, firm and country levels respectively), particularly in the developed economies (Mansell and Wehn, 1998; Oyelaran-Oyeyinka, 2007).

Most of Africa's peoples live in rural areas where agriculture is the main economic activity. However, the productivity and income from agricultural occupations are constrained by such factors as subsistence farming objectives, farmer illiteracy, ignorance and attitudes, small farm sizes, inadequate use of farm equipment and improved seedlings, and low yields. In addition are poor farm gate prices, caused by such other factors as poor processing and storage facilities on farms, and inadequate access to, knowledge of, and participation in, input and produce markets. Hence, poverty continues to pervade rural areas, and it does not require much thought to conclude that dramatic transformation of the growing, processing and marketing processes in agriculture is the needed strategy for mitigating rural and national poverty levels in African countries.

Transforming agricultural processes would require strategies on diverse fronts aimed at removing or lessening the impact of the constraining factors mentioned above. It would entail modernizing agriculture and agro-based activities, including the optimal integration and use of the quintessential productivity enhancing tool of modern societies, ICT, in all stages of agricultural value chains. Hence, the pertinent question: in view of the promise of ICT for transforming productive activities in various sectors, what appropriate mix of ICT with other productive inputs, as well as entrepreneur-behavioural and production process variables would positively transform the productivity of agriculture and agriculture-based activities in African countries?

2. Research problem and objectives

ICT are potential tools for supporting and transforming different aspects of agriculture, from inputs acquisition, through provision or accessing of extension services, farm management, on-farm and off-farm produce processing, to the marketing of produce or processed agro-products. In developed economies, ICT have also been used in research and development to identify improved crop varieties, to access agriculture-relevant environmental and market information, and to develop computerized farm management information systems. But agricultural enterprises and processes in developed and developing countries are worlds apart in terms of scale and other features. Which leads to the important question about the kinds of ICT that are already being used by farmers, produce processors and marketers, to support or transform the agricultural value chain processes in Nigeria and other African countries. Also, what has been the impact of the use of the ICT on the output of the farmers, processors and marketers?

These perspectives motivated this study that investigated the patterns of ICT ownership, investment and use in the value chain of a specific agricultural commodity, cassava, and the possible effects that the uses of ICT have or could have on production process and outcome variables in the value chain. Specifically, the study sought to answer the following research questions, amongst others:

(1) What are the features of the growing, processing and marketing aspects of the cassava value chain in the study area?

(2) What are the levels of and differences in ICT ownership, expenditure and use among growers, processors and marketers in the cassava value chain?

(3) What are the predictors of ICT ownership, investment and use among participants in the value chain?

(4) Do ICT ownership, investment and use bear expected relationships with output values and output-input ratios of the different value chain activities?

(5) What are the relationships between ICT-related variables and input and output prices?

3. Relevant Literature

Agriculture in Developing Countries

Agriculture is the mainstay of the economies of African countries, providing employment and source of food and livelihood for increasing populations (Olagunju, 2007). In Nigeria, for instance, agriculture is the occupation of more than 60 per cent of Nigeria's working population, providing both formal and informal employment for people, among which about 38 percent are females (Opara, 2004). Most farming activities in Africa are undertaken on small scale basis, but the problems of small-scale farmers in Africa are many and multifaceted as explained by Munyua (2007). Laven (2005) and Hess et al. (2007) report that the challenges include: land sub-divisions and small farm sizes, leading to diseconomies of scale and low productivity; inadequate knowledge and skills in modern farming techniques and optimum methods of management; storage difficulties and inadequate farmer experience with the marketing of produce; devastation caused by conflicts, natural disasters and diseases such as malaria and HIV and AIDS; poor linkages between farmers, processors, markets, researchers and extension workers.

ICT in Agriculture and Rural Development

Attempts to understand and improve the effectiveness of ICT investment and use in agriculture have been at the forefront of national and regional policies ever since computers became available for agricultural management and production. ICT investments and applications in agricultural and/or rural activities have payoff potential in research activities (new crop varieties, new planting and processing methods, etc); extension activities (getting information, knowledge and new skills to agricultural firms); production and processing (obtaining and sharing information, preparing operational plans, proposals, budgets, etc); marketing and trade (gaining access to markets, obtaining and using market information); natural resources management and monitoring; rural development and community action.

The uses of digital technologies, especially for communicating agriculture-related information, has been increasing rapidly in developed and emerging market economies. In sub-Sahara Africa, the predominant carriers of information have traditionally been radio, television and newspapers. More recently, mobile phone technologies have diffused very rapidly in the region, providing faster and multi-directional communication and information exchange opportunities for rural dwellers, farmers, extension workers and agricultural firms. But the ICT literacy of farmers in the sub-region is very low, in addition to the absent or negligible Internet connectivity infrastructure in rural areas. Hence, use of computers and the Internet is mostly absent from rural areas.May, Karugia and Ndokweni (2007) observed that one of the key factors affecting use of ICT in agriculture was inappropriate ICT policies especially those addressing rural communities and rural development, language barriers, poor information sharing culture, and the fact that not all people in rural areas have even the low-end ICT such as radios and television.

ICT and Firm Performance

ICT is widely viewed as enabling changes in business and other processes that in turn generate improvements in firms' performance and output. Greater use of ICT may help firms increase their overall efficiency (Pilat,

2004). Other studies have demonstrated that ICT management capabilities, or the managerial skills associated with the acquisition, management and use of ICT have significant impact on business performance (Santhanam and Hartono 2003). Mukhopadhyay, Rajiv and Srinivasan (1997) found support for the beneficial impacts of IT on the quality and output of organizational processes.

It has been argued that ICT investment directly affects production processes and permits other productivity enhancing changes, such as business process redesign and organizational change. ICT investments, particularly those that facilitate sharing of information, may also increase the ability firms to innovate. Sambamurthy, Bharadwaj and Grover (2003) advocated a network of factors that reflect the integration of IT into critical organizational processes and business capabilities. In particular, they conceptualized the organizational impacts occurring through ICT-enabled agility capabilities, digitized organizational processes and knowledge management systems. Barua et al. (1995) proposed a theory of ICT complementarities to argue that the initial effects of IT should occur at the level of organizational processes that use the IT resources.

ICT Use and Impact in Agricultural Firms in Developing Countries

Matambalya and Wolf (2001) studied 300 small and medium enterprises (SME) in Kenya and Tanzania to explore whether use of ICT can enable SME to improve their competitiveness and performance. Their survey found a positive correlation between average enterprise size and use of advanced ICT, which is rising over time. Besides lowering transactions costs, reducing uncertainties and other positive effects well-addressed by most information-theoretic analyses, they concluded that use of ICT can additionally increase competitiveness of SME by facilitating information flows and reducing transaction costs. They also noted that ICT should not be regarded in isolation because it is at best one factor among others that contribute to improved firm performance.

Lubbe (2004) ascertained the linkages between ICT investment and organisational performance in some e-commerce oriented SME in South Africa. The study provides evidence that shows that firm performance is linked to the level of ICT investment intensity in the sample of companies investigated. De Silva and Ratnadiwakara (2008) found that ICT investment reduced the transaction costs of smallholder vegetable farmers in Sri Lanka. Jenson (2007) used micro-level survey data to show that the adoption of mobile phones by fishermen and wholesalers was associated with a dramatic reduction in price dispersion, the complete elimination of waste, and near-perfect adherence to the law of one price.

Adekunle and Alluri (2006) reported on a pilot project that was implemented in South Western Nigeria which provided a multi-purpose community information access point with basic ICT infrastructure including Internet access. The access point provided links to the Community Help Desk and other sources of information on the Internet, provided training to farmers, linked the farmers directly with inputs and output markets, provided an array of farm equipment that farmers lacked on rental basis, and provided online materials to promote farmers' learning of innovative methods to increase their productivity, market products and develop enterprises. Analyses of panel data from the farmers showed that the participating farmers increased their holdings, external inputs used and productivity leading to higher incomes with attendant evidence of graduation into commercial farming.

ICT in Agricultural SME – Actors and Variables

Different studies, including those reviewed above, identify the following as the major categories of agents and variables that influence the output, efficiency and productivity of agricultural SME in developing countries:

(1) Firm characteristics (firm size, as determined variously by farm size, fixed assets, number of workers, working capital, etc; firm age; financial resources, including credit; etc). (Asogwa et al., 2006; Gallacher, 2000; Kassali et al., 2009);

(2) Personal attributes of the owner of firm (sex, age, education, business knowledge and skills, credit-seeking and risk-taking behaviour, etc) (Asogwa et al., 2006);

(3) Entrepreneurial, commercial and innovative orientation of the firm (largely determined, in the case SME, by the personal attributes and business objectives of the owner of firm) (de Mel et al., 2009);

(4) Change agents (extension services and workers, trade and professional associations, technology and input vendors, media, agricultural information systems and networks, etc), which influence the knowledge, perceptions and predispositions of firms and their owners (Adekunle and Alluri, 2006; Al-hassan, 2008; Asogwa et al., 2006; de Mel et al., 2009; Obwona, 2006);

(5) Credit (amounts and terms, as determined by thrift and credit associations, banks, social networks, etc), which influence the scale of operations of the firms in each accounting period (Al-hassan, 2008; Obwona, 2006);

(6) Value chain stage (i.e. whether growing, processing, marketing, etc) of the activity, which determines the corresponding production functions for the activity;

(7) Production technologies, processes and inputs (including farm equipment, seedlings, labour, working capital, etc) actually used by the firms in production activities; and

(8) Prices obtained by actors (particularly growers) in commodity chains for their output which determine their productivity as defined by the ratio of output values (product of output quantities and prices) and input values (product of input quantities and prices).

Research Model

The potential or actual impact of ICT investment and use in agricultural and related activities and value chains need to be evaluated in the context of all these factors or variables, and this is the approach adopted in the case study. However, the following specific variables, which are interrelated in the research model (see Figure 1) were focused upon in the case study:

1. Firm-level (Firm size, Age of firm, Production orientation of firm, Risk behaviour of firm, Technology (ICT) perception of firm, Amount of Credit taken, Change agent visits received);
2. Owner (of firm) level (Sex, Age, Education, Business skills/experience, Risk behaviour, Credit behaviour, Participation in trade and credit associations);
3. ICT Investment and use (ICT ownership, ICT investment, ICT use, perception of ICT usefulness);
4. Production Input variables (including ICT investment). These comprise the quantities, prices and values of inputs;
5. Production Process variables (Communication diversity, communication range, communication intensity; Information processing; Job-related knowledge/skill improvement; Market participation diversity; market participation range, market participation intensity; Research and innovation; Managerial style;
6. Production Output variables. These are the quantities, prices and values of output;
7. Credit availability;
8. Change and agriculture extension agents;

9. ICT availability and accessibility in the environment; and

10. Industry dynamics and competition.

However, due to the small and homogeneous nature of the study area, the data collection and analysis in the case study excluded ICT availability and accessibility in the environment, and industry dynamics and competition, which would be important when comparing value chain activities in larger or disparate geographical areas.

The research model for this study is specified in Figure 1. The research model interrelates firm-owner (i.e. personal and entrepreneurial) with firm-level variables, as well as highlights the potential roles of firm-internal and firm-external factors in determining investment in and actual use of ICT, and the impact of such use on production processes and outcomes.

Figure 1: Research Model

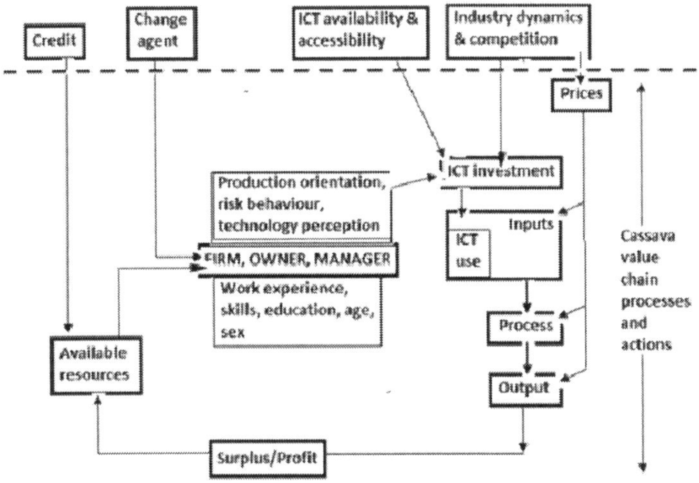

Figure 2: Catalytic and transmission mechanisms among ICT and production-related variables

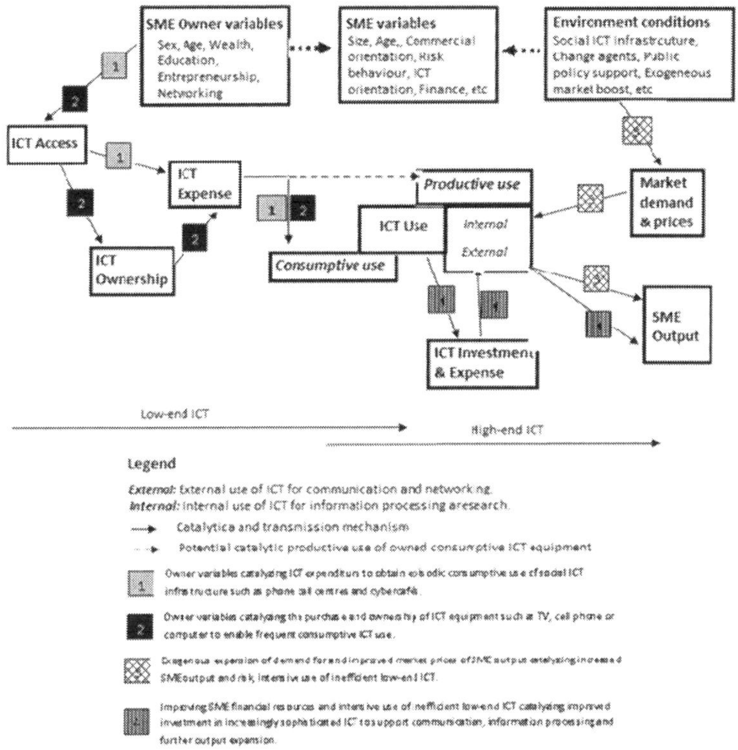

4. Study Environment

Cassava value chain processes and development in Nigeria

Nigeria, with an estimated population of about 120 million (Common Country Assessment, 2001) is the most populous nation in sub Saharan Africa. More than 55% of Nigerians are female, the poverty rate is about 67.8% and about 70% of the people live in the rural areas. Nigeria is the largest cassava producing country in the world, and accounts for 19% of the world output, and 34% of Africa's output (Promoting Value-Adding in Nigerian Agriculture, 2005). By zone, the North Central zone of the country produces about 29% of the national output, while the South South, South West, South East and North West zones produce 24%, 20%, 20% and 7% respectively.

In Nigeria, cassava is predominantly a food crop, compared to Brazil and Thailand where it is mostly an industrial and an export crop, respectively. Nigeria has a large domestic market for cassava, and domestic market prices are good for local industrial development (Promoting Value-Adding in Nigerian Agriculture, 2005). Nigeria does not, however, have a comparative advantage in the export of cassava chips, pellets and starch, due largely to high production cost. Marketing of cassava products as industrial raw material has remained largely unexploited in Nigeria, although post-harvest utilization of cassava has moved to the production of industrial cassava flour (Oyewole and Philip, 2006).

Oyewole and Phillip (2006) explain that traditionally cassava farmers were also the primary processors of cassava produce. Typically, the cassava roots are harvested and processed close to the point of harvest. Cassava is processed into different processed food products that are subsequently prepared into food. The products include garri (which involves the following processing steps: grating into paste, straining, drying, frying, cooking), lafun (cutting into chips, drying, grinding into flour), fufu (soaking, grinding, straining, cooking into ready-to-eat meal), etc.

However, cassava is rapidly being transformed into a cash crop as more industrial and foreign market uses of processed cassava products are being found and promoted by government, including glucose, starch, ethanol, etc. Over time, cassava value chain activities have become increasingly market-oriented and the processing of the bulk of harvested cassava has moved away from farms and growers to processing clusters in nearby large and small towns. Many farmers with large areas of cultivated land now sell their crop to processors or marketers by ready-to-harvest hectares of land, and the processor or marketer performs the actual harvesting for immediate transfer to produce markets or processing clusters. SME in the processing clusters then undertake the processing or hire out processing equipment to third party piecemeal processors most of who are women. Clusters usually specialize in the production of specific cassava products.

A number of initiatives are being implemented by stakeholders, including the governments and cassava growers associations, to increase yields and cultivated area. The Presidential Initiative on Cassava has mandated millers to integrate 10% cassava flour to wheat flour in making bread, a move aimed at increasing the industrial utilization of the crop. Farmers' groups (or clusters) are being organized to use mechanized equipment, high yielding varieties and improved farming practices (Promoting Value-Adding in Nigerian Agriculture, 2005). The International Institute of Tropical Agriculture (IITA), Ibadan, Nigeria is implementing two cassava development projects – the Cassava Chain Development in West Africa (CCDWA) and the Cassava Enterprise Development Project (CEDP).

ICT infrastructure, markets and services in Nigeria

ICT for rural development may be categorized into four main types – radio/TV, telephones, computers, and the Internet. Radio is the most accessible ICT in rural areas, largely because of its portability, low cost and ease of use. Radio programmes in local Nigerian languages targeting farmers and rural dwellers have existed for decades (Omenesa, 1997). In the past few years new government-owned radio stations have been established in several small towns and rural communities across the country, and licenses for private community FM stations continue to be granted. Aderinoye et al. (2007) found that 81.4% (or roughly four out of every five) Nigerians owned a radio, while nine out of every ten Nigerian adults listened to radio. By contrast, the study found that only 31.6% of Nigerians owned television sets. Most of these television set owners are likely to be urban-based. But even then, the epileptic electricity supply situation in Nigerian urban centres constrains the regular use of TV sets.

Access to and use of telecommunications, computers and the Internet in Nigeria gained momentum particularly after 2001 when GSM networks were licensed. The number of fixed lines which was about 700,000 in 2002 had almost doubled by the end of 2005, while the number of mobile lines had risen from about 1.56 million in 2002 to about 20 million by 2007 (Olubamise and Awe, 2007). Since the GSM launch, mobile telephony has rapidly become the most popular method of voice communication in Nigeria. There has been rapid diffusion of cell phone infrastructure and usage in Nigeria since 2001. Road-side call centres now abound especially in urban areas and small towns. These are operated mostly by young women aged 20-29 years who had completed secondary school, although increasing numbers of unemployed male youth are also venturing into that business. Some of them also run shops for the sale of cell phone accessories as a major form of self-employment and livelihood (Obayelu and Ogunlade, 2006).

5. Methodology

A value chain describes the full range of activities required to bring a product or service from conception, through the different phases of production and input of various producer services, to the delivery to final consumers (Kaplinsky and Morris, 2007). A value chain can also be described as a network of labour and production processes whose end result is a finished commodity for sale (Hess et al., 2007). A value chain connects different actors and activities, such as production areas, processing plants or markets. Technology, including ICT, is employed to some degree in every value creating activity, and changes in technology can impact competitive advantage by incrementally changing the activities themselves or by making possible new configurations of the value chain. An agricultural value chain contains several stages that successively add value to an agricultural produce, including stages pertaining to inputs, pre-production, production, harvesting, post-harvest, local value addition, and marketing. The cassava value chain is therefore expected to straddle the various rural and urban communities that participate in the activities of the chain within and/or outside a focal study area/region.

This study focused only on the cassava growing, processing and marketing activities and links in the cassava value chains in Oyo State of Nigeria, which is one of the states in the South West zone of Nigeria where cassava production, processing and marketing is widespread in rural and urban areas and markets. Oyo State is peopled by the Yoruba ethnic group and is culturally homogeneous. The study collected data from three of the five zones of the state – Ibadan, Oyo and Ibarapa – leaving out only the Ogbomoso and Oke-Ogun zones.

The growing, processing and marketing stages in the cassava value chain straddles three different economic sectors – agriculture (growing), manufacturing (processing) and services (marketing). The cassava value chain may be conceived as beginning with the sources of input for cassava production such as markets for the procurement of equipment, seedlings, pesticides, farm labour, etc. Thereafter is the cassava growing stage, where the crop is grown, usually by small holder individuals or SME, although some large farms have emerged recently. The third stage is the produce processing, usually done close to but off the farm. Finally, connecting the cassava growing and processing stages are marketing processes, which are performed by produce or processed product markets and marketers. These intermediate between farms (growers), processing plants, and local, regional and overseas markets for processed cassava.

Data were collected primarily through structured interviews with sampled cassava growers, processors and marketers. The interviews were administered by the researchers and four research assistants. The interviews were based on a comprehensive and structured questionnaire designed to collect both quantitative and qualitative data. The interviews were recorded, except where the interviewees were uncomfortable with the use of the recording equipment.

Preliminary literature and field investigations revealed that most cassava growers, processors and marketers in the study area are registered members of associations of growers, processors and marketers respectively. Growers, processors, marketers who are registered with associations or trade groups are likely to be not too small, and are likely to be market than subsistence oriented. They are also likely to be small and medium enterprises that employed some hired labour, as well as used strategies (including those that are ICT-based) to boost their productivity and output and to seek out new and better input and output markets and prices. The study therefore used the leadership and membership of the associations to identify and select the samples of cassava growers, processors and marketers surveyed in the study.

Snowball and ransom sampling techniques were used to identify prospective interview respondents. The snowball approach was used to identify all the growers, while both methods were used to select processors and marketers. In the snowball approach, resource persons at the International Institute of Tropical Agriculture (IITA), and the Oyo State Ministry of Agriculture, both located in Ibadan, Oyo State, provided information about the leaders of the associations of cassava growers, processors and marketers (traders), and major clusters of processors in and around Ibadan, Oyo/Ilora and Ibarapa/Eruwa cassava growing zones. The leaders of the associations, processing clusters and markets then provided names and addresses of some of their well known members who, in turn, identified other members, and so on.

The random sampling approach was used ad hoc within markets and processing clusters to select some of the processors and marketers. Four research assistants, whose mother tongue is Yoruba, were recruited and trained for the interviews, which were almost all conducted in the Yoruba language. The assistants were trained in the proper translation and asking of questions both during a pre-field workshop and the pilot administration of the interviews that were facilitated by the researchers. The interviews were conducted during April and May 2009, which was during the planting period for cassava timed for harvest in November. One hundred and thirty-nine value chain participants, comprising 56 growers, 39 processors, and 44 marketers, participated in the interviews.

Effort was made, during the design of the interview questionnaire to decompose key variables of the study (e.g. ICT investment, owenership, use, production output, inputs, processes, etc.) into their component variables and questions as necessary. For instance, ICT use was defined in terms of some of the different possible uses of the following ICT equipment and services: (i) radio; (ii) TV; (iii) telephone; (iv) computer; and (v) Internet.

The quantitative analyses included frequencies, correlation and regression. Various simple and multiple linear regression models connecting dependent and independent variables in the research model in Figure 1 were specified, and estimated using ordinary least squares (OLS). In addition, the information obtained from the review of literature, and direct observations during the interviews provided rich context for interpreting the quantitative analyses.

6. Results and Findings

The main findings of the study are summarized, in relation to the research questions, as follows:

1. What are the features of the growing, processing and marketing aspects of the cassava value chain in the study area?

The growing of cassava is found to be dominated by men and processing by women. This is probably an extension of the traditional gender roles played by men and women in the growing and processing of agricultural food crops into processed food or food products. Processing of cassava by small scale enterprises (SME) is usually done with basic processing equipment and is highly labour-intensive. Such work would often be avoided by educated male youth, who might not want to engage in the female-gendered work, particularly if the work is also menial.

Cassava growers are more inclined to seek credit and undertake risky investments than the processors and marketers. Investments in growing activities have longer gestation periods than those in processing and marketing, and are often subject to unpredictable risks from many sources, including weather, pestilence and markets, than processing and marketing. Risk hardiness and credit seeking might therefore be a key requirement for, or an inevitable attribute of, the cassava grower, or farmers in general. By contrast, processors acquire cassava produce from growers or marketers only when ready to undertake processing, as the produce will deteriorate unless processed within a few days. At SME scale of operation, processing is often done piecemeal, and based on actual demand for the processed products. Risk is therefore low. Similarly, marketers acquire cassava produce from the growers only when they have got buyers for them. The processed products, except fufu meal, also store very well, when necessary. Marketers' risks, therefore, depend on their own market analyses and decisions.

2. What are the levels of and differences in ICT ownership, expenditure and use among growers, processors and marketers in the cassava value chain?

Value chain participants owned mainly radio (83.5% of the respondents), cell phones (80.6%) and TV sets (58.3%). The percentage for radio is similar to that reported by Aderinoye et al. (2007) for all Nigerians (81.4%), whilst the percentage for TV was much higher in the present study than that reported by that study for all Nigerians (31.6%). The difference for TV is most probably explained by the location and sample of the present study. South-western Nigeria is among the most literate and economically developed regions of the country. The sample probably contained a higher proportion of wealthier members than one would find among Nigeria's population. Similarly, the finding in respect of slightly more frequent ownership of computers among processors is probably accounted for by a few of the processors who were more educated than most of their peers in the processing business, and who had established cassava processing businesses after exiting or retiring from formal occupations in the urban areas. Such processors might have acquired the computers before starting their processing businesses, but whether they are able to use the computers to process business-related information is debatable.

No significant differences were found between growers, processors and marketers in the purchases of information, expenditure on ICT equipment, and expenditure on ICT-based services. A plausible explanation for this would be that ownership of ICT equipment pertained mostly to low cost radio sets, cell phones and TV sets that are inexpensive to maintain and use, and the use of which is also generally uniform among growers, processors and marketers.

3. What are the predictors of ICT ownership, investment and use among participants in the value chain?

The important predictors of ICT ownership, investment and use among growers, processors and marketers include age of firm, and the firm owner's gender, age, education, knowledge, credit seeking and extent of participation in trade and credit associations. Most of these variables are connectivity and exposure variables, and signpost the need to encourage younger and better educated people to venture into cassava-related businesses, as well as educational programmes and information and business networks that expose the value chain participants to new ideas and innovations, including the innovative use of different types of ICT.

4. Do ICT ownership, investment and use bear expected relationships with output values and output-input ratios of the different value chain activities?

ICT-related variables do not constitute important variables for predicting output values and output-input ratios in the value chain activities. Output value is the product of output quantity and output average selling price. Hence, total output value for a period is the same thing as the turnover of a firm for the period, and is sometimes used to measure the size or intensity of operations, of a firm. The findings indicate that the scale of a firm's operations is directly associated with the ownership of ICT equipment (expected), but not with ICT monthly expense (unexpected) or ICT use (unexpected). Instead, some of the antecedent and production process variables, particularly firm size, communication diversity and intensity, are the significant predictors of output values and output-input ratios in the value chain activities.

5. What are the relationships between ICT-related variables and input and output prices?

In principle, ICT use is likely to help growers both in sourcing for improved inputs and growing methods, and favourable prices for inputs and output. For processors however, ICT use is more likely to help in sourcing for favourable prices for cassava produce and output of cassava products, and less likely to help in the produce processing aspects of their business, as the technology is currently mostly basic and stable. Marketing, being a service activity, provides the widest scope for the profitable use of ICT, because both the core and boundary spanning aspects of marketing entail communication or information processing. This often provides marketers with better access to market information than either growers or processors, and marketers often use this advantage for arbitrage or speculative purposes. The findings of this study suggest that communication diversity and intensity, facilitated increasingly by the ready access to, ownership and use of cell phones, may be additional factors that favour the marketers.

ICT ownership and ICT use helped growers and marketers respectively to obtain better prices for produce or products. However, processors were not so helped by either ICT ownership or ICT use. The findings in respect of growers and marketers support those of Jenson (2007) that the adoption of mobile phones by fishermen and wholesalers was associated with a dramatic reduction in price dispersion and near-perfect adherence to the law of one price. The law of one price is essentially the perfect competition price, which is however likely to be favoured by cassava growers and processors more than marketers, because it assures them that they would not be victims of speculating marketers.

7. Policy Recommendations

ICT use among the cassava value chain participants was limited to basic and low-end ICT, primarily radio, TV and cell phones. Such ICT are only relevant for communicating or accessing oral or transient information, which is usually acquired and used as and when needed, with little effort to store the information for future use. Accordingly, appropriate public policy strategies to promote greater investment in, and use of, more sophisticated

ICT would be those that aim to motivate value chain participants to expand the scale of their operations from small subsistence-oriented sole proprietorships to larger-sized partnerships, collectives and limited liability firms.

Women-dominated processors were the least educated compared to the growers and marketers, and were least inclined to explore markets and engage in frequent market transactions. They were also less likely to own, spend on and use ICT. Thus, there is need for policy strategies that target existing women processors towards enabling them to improve their general and ICT literacy and knowledge through adult education and extension programmes. Also needed are programmes to promote women-controlled cassava processing associations or societies that engage in thrift, credit and collaborative investment, processing and marketing activities. Cassava processing should also be promoted among female secondary school leavers and university graduates who, because they possess higher general education than most of the existing processors, would be more technology and ICT conscious and aim for larger processing operations than the existing processors.

In order to promote use of higher-end ICT in order to catalyse improved output, productivity or efficiency among participants in the cassava value chain, there is a need to stimulate higher investment and scale of cassava growing, processing and marketing activities which would precipitate increased need to use ICT to access and process market or technological information.

8. Conclusion

A fundamental feature of participants in the cassava value chain is the low sophistication of ICT use. Owned and used ICT equipment comprised mainly of radio sets, cell phones and TV sets in that order. These are low-end ICT equipment, which were also used in basic ways: radio and TV mainly to receive news, and cell phones to make and receive voice calls. Sophisticated uses of these equipment for example, radios and TV for advert seeking and placement in respect of inputs and output, or cell phones for fund transfer, SMS messaging, Internet connectivity, among others, were not widespread.

Nevertheless, this study found significant relationships between ICT ownership and output values among processors and marketers. The study also found significant correlations between the ICT-related variables and the prices obtained by growers and marketers for cassava produce and garri product respectively. Taken together with the knowledge of the technical processes of cassava growing, processing and marketing, the findings suggest that ownership and use of ICT equipment (cell phones and radio and TV sets), probably helped value chain participants to obtain favourable prices for their output more than to improve the technical processes of their productive activities. Hence, the correlation between ICT ownership and output values probably derives more from the use of ICT by the participants to identify good markets and favourable prices for output than to improve physical productive processes and output.

The relationships observed in this study between ICT use and output value were however not as strong as expected from theories that predict positive instrumental value of ICT in productive activities. This might be due to the possibility, noted by some studies (e.g. Barua et al., 1995; Sambamurthy, Bharadwaj and Grover, 2003), that ICT use influence production outcomes, not directly, but indirectly through more efficient production processes.

References

Adekunle, A. and Alluri, K. (2006), 'Using ODL aided by ICT and Internet to increase agricultural productivity in rural Nigeria,' Paper presented at the 4th Pan Commonwealth Forum on Open Leaning, Ocho Rios, Jamaica, 30 October - 3 November, 2006. Available at: http://pcf4.dec.uwi.edu.

Aderinoye, R.A., Ojokheta, K.O. and Olojede, A.A. (2007), 'Integrating mobile learning into nomadic education programmes in Nigeria: Issues and perspectives', in *International Review of Research in Open and Distance Learning*, 8(2): 1-17. Available at: http://www.irrodl.org/index.php/irrodl/article/view/347/919.

Al-hassan, S. (2008), 'Technical Efficiency of Rice Farmers in Northern Ghana', *AERC Report RP 178,* April, African Economic Research Consortium, Nairobi, Kenya.

Asogwa, B.C., Umeh, J.C. and Ater, P.I. (2006), 'Technical efficiency analysis of Nigerian cassava farmers: a guide for food security policy', *Paper presented at the 2006 Annual Meeting of the International Association of Agricultural Economists*, 12-18 August, Queensland, Australia. Available at: http://ideas.repec.org/p/ags/iaae06/25473.html.

Barua, A., Kriebel, C.H., and Mukhopadhyay, T. (1995), 'Information technologies and business value: An analytic and empirical investigation,' in *Information Systems Research*, 6(1): 3-23.

De Silva, H. and Ratnadiwakara, D. (2008), 'Using ICT to reduce transaction costs in agriculture through better communication: A case-study from Sri Lanka,' Available at http://www.lirneasia.net/wpcontent/uploads/2008/11/transactioncosts.pdf

de Mel, S., McKenzie, D. and Woodruff, C. (2009), 'Innovative firms or innovative owners? Determinants of innovation in micro, small, and medium enterprises,' *Policy Research Working Paper Series 4934*, The World Bank. Available at: http://www.wds.worldbank.org/servlet/WDSContentServer/WDSP/IB/2009/05/18/0001 58349_20090518125334/Rendered/PDF/WPS4934.pdf.

Gallacher, M. (2000), 'Education as an input in agricultural production: Argentina,' Available at: from http://www.cema.edu.ar/publicaciones/download/documentos/189.pdf (Accessed: 20 October 2009).

Hess, C., Springer-Heinze, A., Bertenbreiter, W., Jarchau, P. and Braun, P. M. (2007), *Services for Rural Development Reader: Pro-Poor Services in Value Chain Promotion – Concept*, Deutsche GTZ Eschborn, Division 45. Available at: http://www.gtz.de.

Jensen, R. (2007), 'The digital provide: Information technology, market performance, and welfare In the South Indian fisheries sector', in *The Quarterly Journal of Economics*, 122(3): 879 – 924.

Kaplinsky, R. and Morris, M. (2007), 'The Structure of Supply Chains and their Implication for Export Supply,' *Paper prepared for African Economic Research Consortium*, Nairobi, Kenya.

Kassali, R., Ayanwale, A.B. and Williams, S.B. (2009), 'Farm location and determinants of agricultural productivity in the Oke-Ogun area of Oyo State, Nigeria,' in *Journal of Sustainable Development in Africa*, 11(2): 1-19. Available at: www.jsd-africa.com/Jsda/V11N02. ../FarmLocationDetermination.pdf.

Laven, A. (2005), 'Towards a Sustainable Cocoa Chain, A Ghanaian Perspective', *Report of a workshop held at the British Council on the 31st of March 2005*, in collaboration with Prof. V.K. Nyanteng and SNV Netherlands Development Organization, Accra, Ghana.

Lubbe, S. (2004), 'The impact of IT investment in RSA e-commerce SME organisations,' in Electronic Journal of Information Systems Evaluation, 7(1): 49-56.

Mansell, R. and Wehn, U. (1998), *Knowledge Societies: Information Technology for Sustainable Development*, New York: Oxford University Press and UNCSTD.

Matambalya, F. and Wolf, S. (2001), 'The Role of ICT for the Performance of SMEs in East Africa: Empirical Evidence from Kenya and Tanzania', *ZEF–Discussion Papers on Development Policy*, Bonn, December. Available at: http://www.zef.de/fileadmin/webfiles/downloads/zef_dp/Zef-dp42.pdf.

May, J., Karugia, J. and Ndokweni, M. (2007), 'Information and Communication Technology and Agricultural Development in Sub-Saharan Africa: Transformation and Employment Generation,' *AERC framework paper for Special Research Project on ICT and Economic Development in Africa*, African Economic Research Consortium, Nairobi, Kenya.

Meier, G. (2001), 'Introduction: Ideas for development,' in: Meier, G.M. and Stiglitz, J.E. (eds.), *Frontiers of Development Economics: The Future in Perspec-*

tive, New York: Oxford University Press and World Bank.

Mukhopadhya, T., Rajiv, S. and Srinivasan, K. (1997), 'Information technology impact on process output and quality,' in *Management Science* (Special issue), 43(12): 1645 – 1659.

Munyua, H. (2007), 'ICTs and Small-scale Agriculture in Africa: A Scoping Study', *Report prepared for the International Development Research Centre (IDRC)*, Ottawa, Canada.

Obayelu, A.E. and Ogunlade, I. (2006), 'Analysis of the uses of information and communication technology for gender empowerment and sustainable poverty alleviation in Nigeria,' in *International Journal of Education and Development using ICT*, 2(3): 45-69.

Obwona, M. (2006), 'Determinants of Technical Efficiency Differentials amongst Small- and Medium Scale Farmers in Uganda: Case Study of Tobacco Growers,' *AERC Report RP12*, February.

Olagunju E.G. (2007), 'Water resources development: opportunities for increased agricultural production in Nigeria'. Available at: http://urn.kb.se/resolve?urn=urn:nbn:se:liu:diva.

Olubamise, B and Awe, J. (2007), 'Nigeria ICT4D Annual Review 2007: A synopsis of the ICT4D sector in Nigeria incorporating activities of the Government, Private Sector and Non-Governmental Organizations'. Available at: http://www.ji daw.com.

Omenesa, Z. E. (1997), 'Rural agricultural radio in Nigeria: An overview of the National Agricultural Extension and Research Liaison Service (NAERLS) Farm Broadcaster,' *Journal of Agriculture Extension*, pp.74-81.

Opara, L.U. (2004), 'Emerging technological innovation triad for smart agriculture in the 21st centurity - Part I: Prospects and impacts of nanotechnology in agriculture,'*Agricultural Engineering International: the CIGR Journal of Scientific Research and Development, Invited Overview Paper*, Vol. VI., July 2004.

Oyelaran-Oyeyinka, B. (2007), 'ICTs and Industrial Development: Transformation and Employment Generation,' *AERC Framework paper for Special Research Project on ICT and Economic Development in Africa*, March, African Economic Research Consortium, Nairobi, Kenya.

Oyewole, O.B. and Philip, B. (2006), 'Agro-food chains and sustainable livelihood: A case study of cassava marketing in Nigeria,' in: Ruben, R., Slingerland, M. and Nijhoff, H. (eds.), *Agro-Food Chains and Networks for Development*, New York: Springer.

Pilat D. (2004), 'Introduction and Summary,' in Economic Impact of ICT, Paris: OECD.

Promoting Value-Adding in Nigerian Agriculture: The Cassava Industry Example (2005), Technical Assistance to the House of Representatives Committee on

Agriculture, *Policy Brief No. 3* (August). Available at: http://Pak-Nigeria.org/Pdfs/15_Ag_Policy_Brief_2-3_Add_Value_Cassava.Pdf.

Sambamurthy, V., Bharadwaj, A.S. and Grover, V. (2003), 'Shaping agility through digital options: Reconceptualizing the role of IT in contemporary firms,' in *MIS Quarterly*, 27(2): 237-263.

Santhanam, R. and Hartono, E. (2003), 'Issues in linking information technology capability to firm performance,' in *MIS Quarterly*, 27(1): 125-153

Acknowledgement

The authors are very grateful to the African Economic Research Consortium (AERC) for funding this research under the ICT and Economic Development in Africa Project.

Chapter 14

Conceptualizing Accountability in ICT4D:

Complementary Perspectives of Information Systems and Development Management

Silvia Masiero

1. Introduction

In this chapter, we conceptualize the academic field of Information and Communication Technology for Development (ICT4D) as constructed by the building blocks of information systems (also known as IS) and development management (also known as DM). On the one hand, the field of information systems focuses on the introduction of ICTs in the sphere of organizations, and, more specifically, on the mechanisms of change and evolution induced by new technologies in the organizational context. This field draws on the fusion between computer science, whose technological insights account for the engineering rationality of digital artefacts, and the psycho-social contents of behavioural science (Avgerou, 2000; Orlikowski and Jacono, 2001; Hevner et al., 2004). On the other hand, development management is centred on institutions, as a key means for achieving progress and empowerment in the developing world: its gnoseological roots are to be found in mainstream institutional theory, which sees institutions as regularized principal-agent relationships, whose non-market input is necessary for imperfect markets to work efficiently (Hirschman, 1970; North,

1995; Rodrik, 2007).

Our argument is that, given the double genesis of the field, the notion of accountability in ICT4D needs to embody both the IS conception, which sees accountability as internal and upward-flowing within organizations, and the DM vision, which constructs accountability as flowing downward to the client, in a fashion that transcends organizational boundaries. Here, accountability in information systems is conceptualized through the model of Roberts and Scapens (1985), which views this notion as the relationship between employers and employees, enforced throughout working life by systems for accounting and reporting. Diversely, accountability in development management is constructed on the basis of Hirschman (1970), who theorizes a three-pronged classification of accounting mechanisms through the ideas of exit, voice and loyalty. Given the complementarity between these two perspectives, it is, in our view, problematic that literature in ICT4D systematically privileges the latter: there is, indeed, a finalistic conception of development orientation, which leads scholars to focus almost solely on the users of ICT4D, black-boxing the internal mechanisms that occur within agent organizations.

As we will see in practice, this asymmetry of attention should be corrected within research. Our fieldwork in Kerala, a state in southern India, focuses on a telecentre project for human development, and shows both dimensions of accountability at work in the dynamics of ICT4D. Through our case study, we illustrate the functioning of accountability toolkits, as conceived by both fields of information systems and development management: the former focuses on accounting within development organizations; the latter is centred on the relationship between the agent and its users. On the basis of fieldwork, we show that IS-accountability, which guarantees intra-organizational cooperation and coherence, directly feeds into DM-accountability, causing a relationship of systematic reinforcement. As a result, ICT4D projects should be examined from both perspectives, and both dimensions of accountability should be granted equal dignity within research.

This chapter is organized as follows. First, we look at the academic structure of information systems and development management, and observe their interplay in the context of ICT4D. Second, we look at the notion of accountability in both IS and DM, and note the DM-accountability bias that the literature presents, due to systematic orientation towards development. Third, we illustrate the case of the Akshaya Telecentre Project, and justify its suitability for illustrating accountability mechanisms: fourth, we discuss this case in the light of both accountabilities, and show their mutual interaction in the on-field context of ICT4D. Finally, we argue in favour of a more equal approach to research, in which both accountabilities are regarded as paramount.

2. Information Systems and Development Management: the Building Blocks of ICT4D

We start this work by positioning ICT4D within the sphere of the social sciences. Our argument here is that, as an academic subject, ICT4D is to be located at the intersection between the fields of information systems and development management, from whose insights it has originated. These fields constitute, in our interpretation, the building blocks of ICT4D, and are both delineated by a set of defining and distinguishing characteristics.

The field of information systems, as of Avgerou (2000: 567), dates back to the early 1970s: its birth can be traced to a set of technology-related studies, which looked at the design of data processing as embedded in the context of organizations. These studies contained the seeds of the IS field, in that they embodied its very primary feature: indeed, instead of looking solely at the technological artefact, they were able to move beyond it, by considering the relationship between the new technology and the organizational transformations that it induced. The object of information systems, indeed, is not the ICT artefact per se, but its operation in the context of organizations, and, in particular, the change that it determines when the new technology is introduced in the organizational sphere. This concept is well-expressed in the work of Hevner et al. (2004), who dichotomize IS literature in the two streams of behavioural and design science: if the latter is predominantly technological, the former is psycho-social in nature, as it theorizes on the reactions of individuals to technological change.

The field of development management, as conceived by mainstream literature, arose around the decade of the 1960s, in concomitance with the first economic studies on decolonization. According to Akpan (2003: 262), this field should be traced back to the ideas of the so-called modernization theorists, i.e. the scholars who sustained that decolonized countries needed to "modernize" in order to achieve economic development. Modernization, in this view, should be sought by a process of imitation of the industrialized West, through increases of sheer economic output, and consisted of a progression through a set of well-codified "stages" of development. The fundamental feature of this model is its utterly economistic nature, which has recently been borrowed by the neoliberal theory of development: indeed, as of Madon (2005: 405), this doctrine views economic growth, achieved through minimization of the state, as the sole determinant of improvements in the people's standards of living.

The deepest challenge to this economistic view of development can be found in Sen (2001), whose idea of development is radically different from the perspective of productivity as a sole key to welfare improvements. Development, in Sen's view, is regarded as a process that increases the freedom of people, seen as the set of capabilities that individuals need for the purpose of achieving their own life plan. In this conception, the welfare effect of economic growth is merely one of several means, for people to remove the "substantial unfreedoms" located on their way: paradigmatically, at the same time, education comes from eliminating barriers to school attainment, and high life expectancy is the outcome of the progressive deletion of roadblocks to health. This multidimensional view of development has been fully adopted by development management, which, indeed, views the accountability of institutions, and not sheer growth of the economy, as the fundamental indicator of progress (Brett, 1999: 18). By doing so, it reflects the general idea that better institutions are conducive to better development (Rodrik, 2007), as they play a proactive role in maximizing people's capabilities.

Yet, what is exactly meant by "institutions" in development management? The seminal work of Hirschman (1970: 5) refers, both simply and precisely, to "regularized relationships between one principal and one agent". Indeed, in the DM view of the world, the poor and underdeveloped are conceptualized as principals, who delegate development projects to several agents: these are actors that, relying on their intrinsic skills and resources, directly engage in developmental problem-solving. An institution is, in this conception, the principal-agent relationship between the two parts, and consequently, development outcomes should improve along with the capacity of agents to skilfully perform their role of delegation.

Table 1: Key Concepts in Information Systems and Development Management

	Information Systems (IS)	Development Management (DM)
Focus	Organizational change induced by the new technologies	Development outcomes gained through institutional improvements
Unit of Analysis	Organizations	Institutions
Academic Features	• Vision of artefacts as embedded in their organizational context • Combination of computing and behavioural science	• Principal-agent theory as the core of institutions • Multidimensional, capability-oriented view of development
Common Denominator	Orientation to practice and concrete problem-solving	

Table 1 summarizes the key concepts in information systems and development management, as they are devised by the literature and embodied in this study. Our argument here is that ICT4D, as an academic field, is proactively constructed by the building blocks provided by the literature in IS and DM. More specifically, these blocks interact with each other, and contribute to delineating the two-pronged sphere that combines ICTs with better development projects.

On the one hand, as Table 1 reveals, these fields present a strong common denominator: namely, their orientation to practice and problem-solving, i.e. their propensity for work on actual issues rather than for abstraction in terms of general theory. On the other hand, differences are prominent between these two spheres, and one of them, namely the diversity in accountability perspectives, lies at the basis of our analysis here. Therefore, the two visions of accountability, which we will examine in detail below, constitute a practical embodiment of the two subjects that lie at the basis of ICT4D.

3. Two Complementary Visions on Accountability: IS- versus DM-Orientation

The concept of accountability is very widely used, for the purpose of project management and evaluation in ICT4D. Accordingly, accountability can be seen as a buzzword in both information systems and development management, as elaborations on this term pervade both literatures in a capillary and systematic way. Yet, the problematization and consequent definition of accountability follows different paths within these two spheres, which lead to very different implications in terms of the analysis of ICT4D projects.

As far as accountability in information systems (henceforth IS-accountability) is concerned, the work of Roberts and Scapens (1985) provides a thorough illustration in terms of its contents. The conceptualization presented here starts from the observation of how, in the world-view of IS, the idea of accountability is shifted away from an abstract, wholly-technical description of accounting practices: what matters is, instead, the set of mechanisms that lead accounting systems to be embodied, through usage, in real-world organizations. This is, in our interpretation, a direct consequence of the nature of IS, which views technological change as inseparable from its organizational context: as such, accountability needs to be seen in context-embedded terms, rather than placed in the domain of theory and abstraction.

In information systems, accountability coincides with "the giving and demanding of reasons for conduct" within organizations (Roberts and Scapens, 1985: 447). The broad and comprehensive nature of this theoretical notion is restricted, in practice, by two limitations. Firstly, limits are imposed on the subject of accountability: in organizations, this coincides with the employees, who are made accountable to their superiors through a codified set of systems of reporting. Secondly, the direction in which accountability flows is specified: it is, indeed, the employer, rather than the subordinate worker, that holds employees accountable, and enforces this relationship through rewards and sanctions (Roberts, 1991). This conceptualization that was originally conceived for private firms, can be easily translated in the domain of the public sector, which relies on specified hierarchies and on the delegation of responsibilities across different levels.

These elements constitute, in our view, a powerful synthesis of the fundamental tenets of IS-accountability. This concept is preoccupied with reasons for conduct, it is seen as internal to the organization, and it flows upward, i.e. from lower-level employees to upper-level employers. The point of intra-organizational accountability is, perhaps, the one that defines IS-accountability in the most specific way: indeed, as of Roberts and Scapens (1985: 448), accountability systems determine the boundaries of the organization, so that they include all its members – and only them. As a result, IS-accountability does not look, per se, at responsibility of the organization towards its users.

The development management concept of accountability (henceforth DM-accountability) is radically different from the IS-dictated one, in all the three respects that we have contemplated. Yet, between these concepts, a similar aspect arises: indeed, in DM as well as in IS, it would be problematic to speak about an abstract idea of accountability, disentangled from the real-life dynamics of organizations and development. This is because, as mentioned above, both fields are practice-oriented in nature, and context-embeddedness is an intrinsic feature of the conceptual toolkit that characterizes both spheres. As such, it is not surprising to find in DM-accountability the very same orientation to practice that we have recognized in its IS symmetrical.

Beyond this point, though, differences between the two fields are significant. Brett (1999: 41) gives perhaps the most comprehensive definition of accountability as confined to DM: namely, "a function of the leverage exerted over agencies by their principals", in which "agencies" are conceptualized as entities that engage in the management of development projects. This means that, naturally flowing from the principal-agent relationship at the basis of institutions, accountability is configured as the extent to which users inform the behaviour of development managers, for them to effectively act in the interest of their beneficiaries. Hence, the situation is turned upside down with respect to IS: here, accountability does not flow from the stronger to the weaker, but from lower levels of power (i.e. the poor and underdeveloped) to higher ones (i.e. development managers). This implies, furthermore, that there is no such thing as identity between systems of accountability and the boundaries of organizations: accountability, as it constructs the objects of development in the role of principals, becomes external in DM, and transcends the intra-organizational dynamics of the material agents of development.

As a result, three symmetric features characterize DM-accountability in practice: it is concerned with the leverage of users in decision-making, it is external to the agent organization, and it flows downward, from development managers to the poor. The institutional theory of development, as codified by Hirschman (1970), identifies three overarching mechanisms, for beneficiaries to exert control on the decisions of agent organizations: firstly, exit resides in the threat, posed by clients, of resorting to another provider – or opting out of the system – if dissatisfied; secondly, voice consists in directly communicating any problems with the agent's work, and informing its improvement by doing so. The third incentive, known as loyalty, relates to what Ebrahim (2003: 814) conceptualizes as "felt responsibility" of organizations: in a set of cases, especially related to the work of solidaristic groups of action, organizations retain their members through faithfulness to a mission, and they have an incentive to preserve their customers by behaving coherently with their own set of values. Exit, voice and loyalty are, therefore, the key mechanisms through which leverage of development beneficiaries is operationalized into actual pro-poor arrangements.

Table 2: Key Features of IS- and DM-Accountability

	IS-accountability	DM-accountability
Definition	The giving and demanding of reasons for conduct	Function of leverage of beneficiaries on their agents
Focus	Internal to organizations	External to development agents
Direction	Upward – from employees to employers	Downward – from development managers to beneficiaries
Mechanisms	Intra-organizational systems for accounting and reporting	Exit, voice and loyalty

Table 2 summarizes the defining features of IS- and DM-accountability, in comparison with each other. The request of reasons for conduct is matched by the requirement of leverage to the poor; intra-organizational accountability to employers is responded by external accountability to the objects of development; upward flows of reporting correspond to downward accountability of the managers of development. The question here is, how does the literature in ICT4D engage with these clashing conceptualizations of the same term?

The problem, as identified within the existent literature, can be stated as follows: current analyses of projects in ICT4D tends to prefer, as if there were no alternatives, the DM-oriented concept of accountability, stressing responsibility of agents towards external beneficiaries rather than their own intra-organizational dynamics (Gasco, 2003; Jaeger and Thompson, 2003; Bhatnagar, 2004; Prakash and De, 2007; Bhatia et al., 2009). Indeed, to our knowledge, no work so far has dealt explicitly with the trade-off between internal, upward accountability within development agencies, and external accountability to the recipients of ICT4D projects. The overarching preference towards DM-accountability has, in our view, a quite precise rationale in ICT4D: indeed, in this field, technologies are subordinated to the purpose of development, and projects are seen and evaluated in the finalistic, beneficiary-related vision that this purpose entails.

That being said, there are, as we submit, strong reasons for scholarship to put this finalistic attitude into question. Firstly, by focusing entirely on external accountability, this vision neglects the dynamics that occur at the various levels of the intra-organizational sphere. This leads, in effect, to black-box all the facts and mechanisms within the back-end side of ICT4D projects, so that development outcomes from new technologies cannot at all be traced to their organizational dynamics. Secondly, as we will see in the discussion of the case study, it might be that IS- and DM- accountability are mutually interacting, or, more specifically, that the former directly feeds into the latter: indeed, organizations can hardly be accountable to their customers, if internal efficiency is undermined by lacking attributions of responsibility.

As a result, the attitude of the literature that focuses mainly on accountability to the clients is put into question by this work. To illustrate this point, the case study of an established ICT4D project in southern India is selected and analyzed using f IS- and DM-accountability framework. From a theoretical viewpoint, this helps to fully recognize the double scientific matrix – resulting from the interplay of IS and DM – that influences academic work on the projects of ICT4D. From a practical position, however, this will uncover the underlying mutual interaction between IS- and DM-accountability, and will serve as a basis to highlight implications in terms of project analysis and research.

4. Data Collection: Methodology and Fieldwork

Data collection for this study has been performed during two visits to the state of Kerala, conducted in summer 2009 and autumn 2010 respectively. The first one was to Malappuram district, the administrative unit where the pilot implementation of Akshaya, the telecentre project observed here, was originally led. This visit has involved a major focus on the activity of local telecentre entrepreneurs, as fieldwork was led in close cooperation with the Akshaya Project Office, operating in Malappuram at the district level. Diversely, the second visit was conducted in the district of Thiruvananthapuram, the capital city of Kerala, where government bodies at the state level are located: here, we have interacted directly with government agencies involved in the project, focusing on the drivers determining layers of power and accountability within the Akshaya initiative. Interviews have been, therefore, conducted across multiple layers of responsibility and influence, and have been systemically complemented by participatory observation in telecentres: in this respect, we have used triangulation to mediate between participant observation and interviews, as suggested by Atkinson and Coffey (2003: 420-422).

One caveat here is that drawing on a single case study, when inferring general lessons for a broader contextual area, is implicitly problematic, given the limitations of single cases in term of spatial and demoscopic coverage (George and Bennett, 2001). Yet, with respect to our Kerala fieldwork, we sustain that our methodological choice is to be justified by two considerations. Firstly, our research here does not aim at drawing policy implications for telecentre managers, a task that, given the complexity inherent in drawing business plans for telecentres, constitutes a challenge in itself (Oestmann and Dymond, 2001): rather, we use the Akshaya project to grasp a mechanism in action, for providing a paradigmatic picture of accountability channels in ICT4D. Secondly, rather than being a single-site, one-off experience, Akshaya is a state-wide project, whose telecentres are spread across all the fourteen districts in Kerala: as a result, the project is a polymorphous venture, whose articulation over a plethora of fields makes it interpretable as a multiple experiment of ICT4D.

The method that we have used to systematize the data gathered on field, in order to make the link between the case-study and the theoretical arguments set out in this study, coincides with the one that Burawoy (1998: 16), in his theorization of reflexive science, refers to as "reconstruction" of theory. In order to produce general conclusions, this method, instead of focusing on the choice of a representative sample, aims to generate knowledge starting from the empirical data that fieldwork produces. This means that we have started our work on the ground of our preferred theoretical assumptions, and, as fieldwork went on, we have modified and shaped our theoretical insights on the basis of the findings that we have come across. This process has enabled us to provide a detailed account on the functioning of accountability mechanisms in Akshaya, from both a development management and an information systems point of view.

5. Akshaya: Telecentres and Development in a Public-Private Interplay

Our case study, located in the southern-Indian state of Kerala, is focused on telecentres, which are broadly defined in the literature as all the "shared premises where the public can access information and communication technologies" (Roman and Colle, 1999: 1). These are, in other words, physical centres that provide local communities with access to the global network, in the form of telephones, computers, the Internet and other ICT-related devices. In telecentres, a franchising model of operation can arise: in this pattern, the state supervises the project and provides Internet-based services through it, but the single telecentres are left in the hands of entrepreneurs, which look after their financial and social sustainability.

The Akshaya Telecentre Project, which is grounded on this public-private model, originated in 2002, from a proposal of the Gram Panchayat – local administrative council – of Malappuram, a district in the north of Kerala. The ICT-dedicated government agency of Kerala – the Kerala State Information Technology Mission (KSITM) – took up the proposal, and implemented it by creating a network of e-kiosks in Malappuram: these centres, managed by selected members of the local community, had the purpose of running a program of computer education (e-literacy), in which one member of each local household would be invited to participate. In the first phase of the project, KSITM planned the geographic location of 630 e-kiosks, selected the entrepreneurs to manage them, and sustained 85% of the e-literacy program's cost to the entrepreneur (Pal et al., 2006: 291).

Performance analyses on the pilot-project phase of Akshaya reported impressively high scores in terms of the objectives attained: as of Madon (2005: 408), the project achieved 100% household e-literacy in Malappuram district. As a result, at the end of the e-literacy period, Akshaya telecentres remained in operation, presenting two major differences with respect to the pilot phase. Firstly, government financing through KSITM was confined to the initial stage: after that, it has been up to telecentre entrepreneurs to run the kiosks, on the basis of their own skills in terms of business and social development. Secondly, the e-literacy programme has been substituted by a new host of applications: citizens can now attend advanced courses of computer training, access applications for trading their goods online, get information on current market prices, and benefit from a suite of e-governance toolkits, which range from the payment of bills to the filing of tax returns.

One caveat, to be made explicit as we ground our analysis on Akshaya, is in terms of the context of this project, which is extremely sui generis within the developing world. The development experience of Kerala, indeed, is often referred to as a "development paradox" (Chopra, 1982; Subrahmanian, 1990; Gopakumar, 2007), because it combines very high levels of human development, especially in terms of literacy and life expectancy, with very low levels of per capita GDP (Government of India Planning Commission 2005: 60). Kerala is, indeed, the only Indian state where the Communist Party of India has enjoyed electoral success since its inception, proposing a model that relied crucially on political awareness and economic redistribution. As a result, sustained public action has conducted to high levels of human empowerment (Parayil, 2000), whereas the overall imperative of redistribution has impeded to reach high rates of economic growth, and ultimately led to a fiscal crisis that constrained the capacity and potential of state-based welfare programmes (George,1993; Tharamangalam, 1998).

Figure 1: Levels of Local Government in Kerala

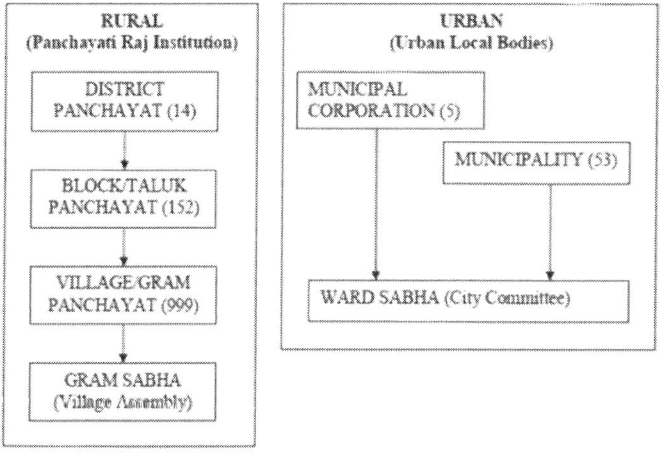

Source: Adapted from Venugopal and Yilmaz (2009)

This model of development, where public action arises as a key feature of citizenry, has one key implication in terms of our analysis, as we focus on the accountability of a state-directed project. Indeed, the political tradition of Kerala has been grounded crucially on the transparency of institutions, achieved through the participation of citizens to the mechanisms of local government: this is why administrative decentralization – activated in India by the 73rd Amendment to the Constitution – has been taken very seriously by the state, so that very high levels of responsibility and financial discretion have been delegated to the district, block, and village panchayats in the countryside, and to their urban counterpart – municipalities and municipal corporations (Véron, 2001; Heller, 1995). Layers of local governance, enacted by the Panchayati Raj reform in 1993, are reported in Figure 1 as of their Keralite configuration: this shows the multiple strata of power that have been introduced in the state, in a project of reform that uses decentralization as a carrier of decisional power to citizens.

The participatory nature of institutions is, indeed, very helpful in terms of our illustrative study, as it facilitates the task of opening the "black box" of accountability mechanisms, flowing between the diverse tiers of the Akshaya project management. Yet, by our analysis, we do not aim to generalize, or to assume that the condition of Kerala can be automatically extended to the developing world. This case, in effect, has been selected exactly because of its peculiar transparency, which is conducive to the illustration of accountability flows in ICT4D.

6. IS- and DM- Accountability in Action: Mechanisms, Instruments and Interdependence

Our discussion of the Akshaya project, led in the light of accountability as conceptualized by both IS and DM, starts from the acknowledgement of the bias that we have previously recognized in the literature: namely, ICT4D projects tend to be examined from the standpoint of DM-accountability, neglecting the sphere of IS-accountability as enacted within development programmes. Scholarship on the Akshaya project is, in effect, paradigmatic of this attitude. The existing literature on Akshaya can, indeed, be clustered into three main bodies: the first one focuses on the state's accountability to citizens-clients (Gurumurthy et al., 2005; Pal et al., 2006; Pal, 2009); whereas the second picks entrepreneurs, and their capacity of trust-building within the local community, as the main unit of analysis (Kuriyan et al., 2006; Gopakumar, 2007).

Diversely, the third body of literature does not explicitly mention accountability issues: yet, it is relevant in that it conceptualizes the project's success not in terms of internal functioning, but solely in terms of acceptance and participation of citizens to the project (Antin, 2005; Kortemann, 2005). The one exception in the literature so far is constituted by Madon (2007, 2005 and 2004), who carried out on-field observation of Akshaya at several points in time: as such, her work clearly illuminates the relationships between actors, and their evolution as the project has proceeded over time. Our work actually builds on that of Madon, as we focus on the specific cluster of relationships that ascribe to accountability, within Akshaya and in the linkages of the project to its recipients.

Our purpose, in this discussion, is that of filling the existent gap in terms of analyses of IS-accountability. Hence, we first look at the internal mechanisms of accountability that flow within the Akshaya project, from the lower-layer of telecentres to the upper-layer of state level management. Second, we look at DM-accountability to users – conceptualized in terms of exit, voice and loyalty – and make explicit the mechanisms through which IS-accountability directly feeds into its acclaimed success. Interdependence between these two dimensions constitutes the practical illustration of the argument that we have stated in theory: namely, ICT4D projects need to rely on both sides of accountability, for yielding a sustainable and continued rate of success.

Figure 2: IS-Accountability Flows in the Akshaya Project

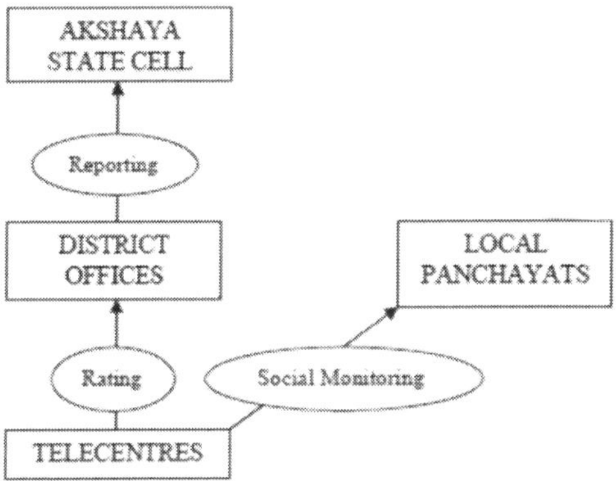

Figure 2 illustrates IS-accountability as it is enacted within the Akshaya project. As shown by the arrows schematized above, telecentres are held accountable to district offices, that directly monitor the project's implementation, and to panchayats, whose political supervision involves control over the work of e-kiosks. Furthermore, district offices are held accountable to the Akshaya State Cell, which embodies the guidelines dictated by KSITM and is, in the Akshaya scheme, the highest expression of government dictates. The mechanisms that operationalize IS-accountability into practice need to be disentangled and analyzed one by one.

Firstly, telecentre entrepreneurs are held accountable by district offices. These offices rely on rating, i.e. the yearly evaluation of each single e-kiosk, as the primary mechanism of accountability. Rating procedures from the district offices are well-codified and enacted: firstly, a set of critical dimensions is established, ranging from financial viability through revenues to effective social impact through participation; then, every year, dedicated officials physically visit each telecentre, in order to establish a scale-based judgement (from A to E) for each dimension and for the e-kiosk as a whole. Fieldwork in Malappuram has revealed that entrepreneurs, as such, are very concerned with rating, because evaluation directly conditions the degree of reputation of their e-kiosks; furthermore, the very design of Akshaya – in which the viability of telecentres is totally up to entrepreneurs – constitutes an incentive for them to work proactively, in order not to lose their critical mass of clients.

Secondly, local panchayats – which are not directly involved with Akshaya, but need to integrate the project within local administration – also play a supervising role in this respect. This was already evident in the Malappuram pilot project, where members of local panchayats worked with KSITM in the selection of telecentre entrepreneurs (Gurumurthy et al., 2005:144): still, after the state-wide rollout, control of panchayats on single entrepreneurs has been largely and firmly maintained. Although panchayats lack the direct incentive of rating procedures, entrepreneurs report to them in terms of the social initiatives, which are led within the telecentres for maximization of community empowerment (Madon, 2007): examples, in this respect, include the proactive involvement of women, and the utilization of telecentres as meeting hubs for NGOs.

Thirdly, district offices are held accountable to the Akshaya State Cell, a coordinating body presided directly by the director of KSITM. This body, which makes the link between the state government and Akshaya local units, supervises activity at the district level, acting as a tool for transmission of KSITM's insights into district entities. Two systems of reporting are at work at this stage: the first one, which is backward-looking, consists of an accounting device that reports on the health status of telecentres in each district, based on the rating that is previously conducted on the activity of all the single e-kiosks. The second tool, whose nature is forward-looking, consists of another reporting mechanism, which contains the entrepreneurs' feedback on telecentre activity, their complaints, and their constructive suggestions for proactively improving upper-level project management.

Therefore, a three-pronged monitoring structure is at work within the Akshaya project. District offices, through the mechanism of rating, exert direct control on entrepreneurs; panchayats, through the monitoring of social initiatives, provide completion of this assessment; the Akshaya State Cell, through both backward- and forward-looking devices, controls the project at the level of district units. This organizational toolkit presents, in practice, the three features that we have noted in our conceptualization of IS-accountability: indeed, it is internal to the development agent (i.e. the

Akshaya project staff), it moves upward (from telecentre entrepreneurs to the Akshaya cell, passing through district-level monitoring); and it leads entrepreneurs and district offices to provide reasons of conduct in terms of project management.

Figure 3: DM-Accountability Flows in the Akshaya Project

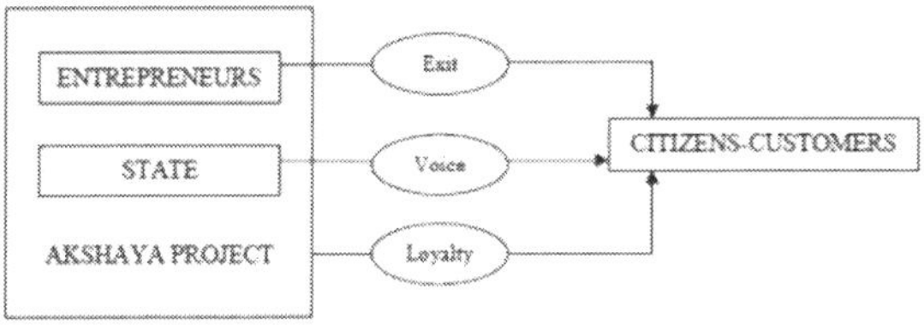

In the Akshaya project, also DM-accountability is organized on a three-level scale, which is illustrated in Figure 3. The first level is that of entrepreneurs, which need to be made accountable to telecentre users, as frontline providers of Akshaya services: at the second level, the state is held accountable by citizens-customers, for the coordination and monitoring of the project as a whole. Thirdly, there is another level of accountability, which is embodied by the project per se, as a synthesis of public and private operations: Akshaya, indeed, was created in the pursuit of a precise set of social objectives, to which the project needs to stick as time goes on, and business-related temptations come about at various levels.

As far as entrepreneurs are concerned, accountability to users works through exit, i.e. through the awareness that, if the service provided by an entrepreneur is not adequate, clients will opt out of her e-kiosk, by moving to another one. Threat of exit, indeed, is enforced by the geographic planning of the Akshaya project: indeed, every house in Kerala is, at maximum, at a 2 km distance from the closest telecentre, a capillarity that was already embedded in the pilot project in Malappuram (Madon, 2005: 408) and that makes it a credible threat that, if dissatisfied, clients will resort to the closest competitor. We can observe, more generally, that threat of exit is strengthened by market incentives, as they are embedded in the Akshaya project design: indeed, in order to attract and retain a critical mass of users, entrepreneurs need to play an extremely proactive role, by continuously liaising with their clients and adapting service provision to their needs. If entrepreneurs are not good enough in these operations, their e-kiosks will not achieve financial viability, which will lead their ventures to failure and ultimately to shutting down: the massive rate of closure, observed in Malappuram during the pilot project, is a concrete proof of the actuality of this danger.

The state, embodied by district offices and by the Akshaya State Cell – as well as by KSITM, the project's leading agency and coordinator – is made accountable to the citizens through voice. Indeed, not only are citizens constantly encouraged to express their feedback on telecentre services, as homogeneously reported by interviews with users in Malappuram: most relevantly, customers' feedback is materially operationalized into the provision of new services, conceived on the basis of the demand of users. This is the case, for example, of railway ticketing, that has been made available through Akshaya as a consequence of the collective request of clients. To understand why the voice mechanism works so efficiently here, it must be noted that Kerala thoroughly retains two of the factors that, in the analysis of Paul (1992: 1050-1051), are configured as primary voice-enablers: firstly, the absence of legal barriers to people's militancy, given that Kerala – on the contrary – strongly relies on public action as an integral part of its development model. Second, the high rates of education among citizens, which lead them to abate the barriers of informational asymmetry, and to codify their demands in a clear and efficient way. The push factor of good organization in the civil society, and the state's receptivity of the citizens' demand pull are the two key elements that interact in the achievement of voice-based accountability of state agencies.

Finally, the project as a whole – as an organic unity composed by entrepreneurs, state coordinators, and their mutual relationships – needs to maintain the requirement of accountability to its objectives. Indeed, to preserve the loyalty of customers, and its credibility as government-branded, Akshaya needs to stick to the goals of social development, and of people-wide gains through connectivity, for the pursuit of which it was originally designed. The problem, in this respect, is raised by the work of Kuriyan et al. (2006): in their analysis, social orientation is threatened by a general veer towards business objectives, brought about by the very same market incentives that make exit an actual threat for telecentre entrepreneurs. The point here is that, as argued in a previous account of our fieldwork (Masiero, 2011), social sustainability in Akshaya e-kiosks does not exclude financial viability: on the contrary, increased participation of people – achieved through trust-building, context-based services, and civil society involvement – reinforces financial outcomes, by attracting and retaining a critical mass of telecentre users.

As a result, in the picture provided, Akshaya arises as an harmonious combination of exit, voice and loyalty, where each dimension is applied to a specific level of DM-accountability. Entrepreneurs are held accountable by the threat of exit; the state informs its action to the citizens' voice, and the need for customers' loyalty prevents business objectives from prevailing over social development goals. This picture reinforces the coherence of Akshaya with the context of Kerala, a state where public orientation leads to trust towards government initiatives, and to general distrust towards business-ended mechanisms (Tharamangalam, 1998; Madon, 2005: 410). The DM-accountability portrayed here leads us to see the functioning of the pattern stated in theory: it is external to the development agent; it flows downward towards citizens-customers, and it has the purpose of maximizing the leverage of users on project management.

Yet, could this harmonious combination of DM-accountability mechanisms be achieved, if IS-accountability was not in place at the internal level? It is, now, our task to uncover the channels of interaction between the two dimensions of accountability, as we have identified them within the Akshaya project. The bottom line of my argument is as follows: there are two mechanisms, flowing from IS- to DM- accountability, which lead the latter to depend crucially on the former, and to be, in effect, hardly achievable without it. Indeed, whereas the threat of exit may function independently as a consequence of market incentives, voice of citizens and loyalty to key objectives could not be operationalized into practical arrangements, if intra-organizational accountability was not effectively put into place.

Figure 4: Mechanisms of Interaction between IS- and DM-Accountability

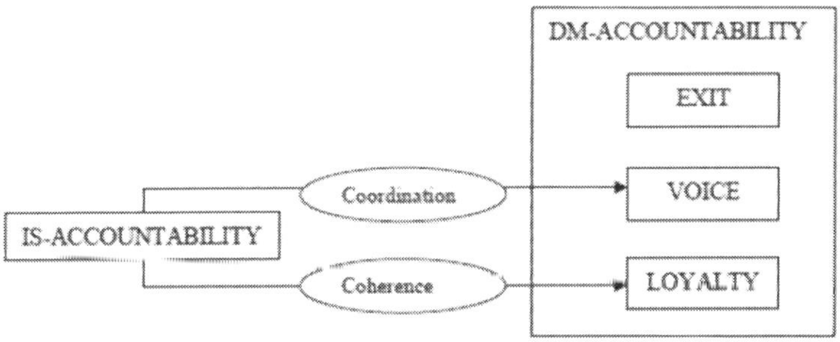

The channels of interaction between IS- and DM-accountability are illustrated in Figure 4. Firstly, IS-accountability provides the Akshaya project with a degree of internal coordination that, accounting for communication between implementing agencies, is instrumental for the voice of citizens to work effectively. Indeed, when customers present their collective demands to the state, frontline providers need to respond by adapting their own system of services: as such, requests are made to KSITM, and need to flow smoothly through the Akshaya State Cell and district offices, before reaching the stage of telecentre-based implementation. Therefore, when KSITM is faced with a new proposal (as it has been, for example, for railway ticketing), this needs to be passed efficiently through lower-level stages of action, which is only possible in force of the IS-accountability mechanisms outlined above.

Secondly, IS-accountability, which consists of a three-level monitoring – through district offices, panchayats, and the Akshaya State Cell – is inherently instrumental in maintaining the project's coherence, i.e. the basis of customer loyalty. This is because reciprocal monitoring between implementing agencies, and the political control of panchayats over telecentre activity, are functional in preventing business temptations from being followed by entrepreneurs: if a part of the project deviates towards business, the rest of the organic unity of Akshaya is there to flag the situation, and correct it for social outcomes to be righteously pursued. Internal coherence is, in effect, a direct output of internal accountability mechanisms: as telecentres directly report to district offices, and are supervised by panchayats and by the Akshaya State Cell, IS-accountability provides concrete arrangements for maintaining Akshaya's unity in enacting its mission statement.

As a result, on the one hand, the exit mechanism is a given in Akshaya, as it is the output of the market incentives that the project, peculiarly in ICT4D, embodies in itself. But voice and loyalty, which need strong and efficient interaction between the project's parts, could not exist without a thorough, well-planned functioning of IS-accountability, which has been overlooked by the literature as a consequence of the finalistic bias towards the DM aspect. Our analysis of the Akshaya project has shown that, in fact, this bias implies neglect of intra-organizational reporting, which might be the very cause of efficiency in development management: as such, one-sided analyses give at best a partial view on ICT4D, which forgets the IS-related building component of the projects. Privileging DM-accountability can be functional to development-oriented analyses, but it leads scholars to black-box a fundamental part of the process: as such, equal dignity should be recognized to both dimensions, as embedded in the dynamics of research.

7. Conclusion

In this chapter, we have conceptualized ICT4D as an academic field constructed by the two building blocks of information systems and development management. We have argued that, given this double genesis, the notion of accountability in ICT4D should embody both the IS conception of this term, which is internal and upward within organizations, and the DM vision of the same concept, which, flowing towards clients, is external and downward. Our fieldwork in Kerala, on a public-private telecentre project for development, has allowed us to illustrate the operational mechanisms of both accountabilities; as such, we have shown that DM-accountability, which is privileged in the literature, is constructed on the grounds of internal cooperation and coherence, which are instrumentally provided by IS-accountability. As such, we conclude that projects in the field of ICT4D should be examined from both perspectives, ensuring that IS- and DM-accountability are granted equal dignity within the research process.

Implications for further research stem directly from this conclusion. As we problematize the DM-oriented bias in the study of accountability within ICT4D, we suggest that more research is conducted in the field of IS-accountability, i.e. on the internal systems of accounting and reporting that occur at the internal level of development organizations. Opening the black box of IS-accountability constitutes, first of all, correct scientific conduct, as it recognizes the dependence of ICT4D on information systems as a building-block discipline. Furthermore, this operation is useful in practice, due to the fact that, as the Akshaya case has shown, DM-accountability can be tied to the IS concept through a direct relation of cause and effect.

Finally, in a broader fashion, a memo for ICT4D researchers should include thorough reliance on IS and DM as the building blocks of the subject. This double genesis that we have illustrated here in the light of accountability, touches on several dimensions of assessment and evaluation, and should be kept in mind when approaching ICT4D. As such, artefacts should not be tackled from the viewpoint of innovation per se: rather, what matters is their capacity of effecting organizational change, and of translating it into outcomes in terms of development on institutional grounds. Moreover, and most importantly, development outcomes outstrip the field of productivity-based economic growth: ICT4D, as a polidimensional sphere, should be sought in all aspects of people's empowerment, in a principal-agent fashion that transcends economistic models.

References

Akpan, P.I. (2003), 'Basic-Needs to Globalization: Are ICTs the Missing Link?', in *Information Technology for Development*, 10(4): 261-274.

Antin, J. (2005), 'The Case for Culturally Appropriate Kiosks: an Ecological Approach to Technology and Culture', in *Annual Meetings of the American Anthropological Association*, November, Washington, D.C.

Atkinson, P. and Coffey, A. (2003), 'Revisiting the Relationship Between Participant Observation and Interviewing', in Holstein, J. A. and Gubrium, J.F. (eds.), *Inside Interviewing: New Lenses, New Concerns*, London: Sage Publications.

Avgerou, C. (2000), 'Information Systems: What Sort of Science Is It?', in *Omega*, 28(5): 567-579.

Bhatia, D., Bhatnagar, S.C. and Tominaga, J. (2009), 'How Do Manual and Government Services Compare? Experiences from India', in Khalil, M., Dongier, P. and Zhen-Wei Qiang, C. (eds.),*Information and Communications for Development: Extending Reach and Increasing Impact*, Washington, D.C.: World Bank Publications.

Bhatnagar, S. (2004), *E-Government from Vision to Implementation: a Practical Guide with Case Studies*, New Delhi: Sage Publications.

Brett, E. A. (1999), 'Understanding Institutions and Organisations', in Harriss, J., Hewitt, T. and Robinson, D. (eds.), *Managing Development: Understanding Inter-Organizational Relationships*, London: Sage.

Burawoy, M. (1998), 'The Extended Case Method', in *Sociological Theory*, 16(1): 4-33.

Chopra, P. (1982), 'The Paradox of Kerala', in *World Health Forum*, 3(1): 74-77.

Ebrahim, A. (2003), 'Accountability in Practice: Mechanisms for NGOs', in *World Developmen*t, 31(5): 813-829.

Gasco, M. (2003), 'New Technologies and Institutional Change in Public Administration', in *Social Science Computer Re*view, 21(6): 6-14.

George, A. L, and Bennett, A. (2003), *Case Studies and Theory Development in the Social Sciences*, Cambridge: Harvard University Press.

George, K. K. (1993), 'Limits to Kerala Model of Development: An Analysis of Fiscal Crisis and its Implications', *Monograph Series*, Trivandrum: Centre for Development Studies.

Gopakumar, K. (2007), 'E-Governance Services Through Telecentres: the Role of Human Intermediary and Issues of Trust', in *Information Technologies and International Development, 4(1): 19-35.*

Government of India, Planning Commission (2005), Kerala Human Development Report, Trivandrum: Independent Publication Group.

Gurumurthy, A., Parminder, J. S. and Gurumurthy, K. (2005), 'The Akshaya Experience: Community Driven Local Entrepreneurs in ICT Services', in O'Siochru, S. and Girard, B. (eds.), Community Based Networks and Innovative Technologies, New York: United Nations Development Programme (UNDP).

Hevner, A. R., March, S. T., Park, J. and Ram, S. (2004), 'Design Science in Information Systems Research', in *MIS Quarterly*, 28(1): 75-105.

Heller, P. (1995), 'From Class Struggle to Class Compromise: Redistribution and Growth in a South Indian State', in *Journal of Development Studies*, 31(5): 645-672.

Hirschmann, A. O. (1970), *Exit, Voice, and Loyalty: Responses to Decline in Firms, Organizations, and States*, Cambridge: Harvard University Press.

Kortemann, M. (2005), 'Cultural Background and Technology Acceptance: Evaluation of ICT Projects that Bridge the Digital Divide', in Day, D. L., Evers, V. and Del Galdo, E. (eds.), *Designing for Global Markets 7, Proceedings of the Seventh International Workshop on Internationalisation of Products and Systems.*

Kuriyan, R. Toyama, K. and Ray, I. (2006), 'Integrating Social Development and Financial Sustainability: the Challenges of Rural Computer Kiosks in Kerala', in *Proceedings of the ICT for Development Conference*, Berkeley, May.

Jaeger, P.T. and Thompson, K. M. (2003), 'E-Government Around the World: Lessons, Challenges, and Future Directions', in *Government Information Quarterly*, 20(3): 389-394.

Madon, S. (2007), 'Telecentres and Development: a Social Space Approach', *Working Paper No. 164*, Information Systems and Innovation Group, London School of Economics and Political Science.

Madon, S. (2005) 'Governance Lessons from the Experience of Telecentres in Kerala', in *European Journal of Information Systems*, 14(4): 401-417.

Madon, S. (2004), 'Evaluating the Developmental Impact of E-Governance Initiatives: an Exploratory Framework', in *The Electronic Journal of Information Systems in Developing Countries*, 20(5): 1-13.

Masiero, S. (2011), 'Financial versus Social Sustainability of Telecentres: Mutual Exclusion or Mutual Reinforcement?', *Electronic Journal of Information Systems in Developing Countries*, 45(3): 1-23.

North, D. (1995), 'The New Institutional Economics and Third World Development', in Harriss, J., Hunter, J. and Lewis, C.M. (eds.), *The New Institutional Economics and Third World Development*, London: Routledge.

Oestmann, S. and Dymond, A.,C. (2001), 'Telecentres - Experiences, Lessons and Trends', in Latchem, C. and David Walker (eds.), *Telecentre: Case Studies and Key Issues*, Vancouver: Commonwealth of Learning.

Orlikowski, W. and Jacono, C. (2001), 'Research Commentary: Desperately Seeking the 'IT' in IT Research - a Call to Theorizing the IT Artefact', in *Information Systems Research*, 12(2): 121-134.

Pal, J. (2009), 'If the State Provided Free Computer Literacy, Would It Find Takers? Evidence and Propositions from the Akshaya Project in India', in *Information Systems Frontiers*, 11(2): 105-116.

Pal, J., Nedevschi, S., Patra, R. K. and Brewer, E. (2006), 'A Multidisciplinary Approach to Studying Village Internet Kiosk Initiatives: the Case of Akshaya', in *E-Learning*, 3(3): 291-316.

Parayil, G. (2000), 'Introduction: Is Kerala's Development Experience a 'model'?', in Parayil, G. (ed.), *Kerala - the Development Experience*, London: Zed Books.

Paul, S. (1992), 'Accountability in Public Services: Exit, Voice and Control', in *World Development*, 20(7): 1047-1060.

Prakash, A. and De, R. (2007), 'Importance of Development Context in ICT4D Projects: a Study of Computerization of Land Records in India', in *Information Technology & People*, 20(3): 262-281.

Rodrik, D. (2007), *One Economics Many Recipes: Globalization, Institutions, & Economic Growth*, New York: Princeton University Press.

Roberts, J. (1991), 'The Possibilities of Accountability', in *Accounting, Organizations and Society*, 16(4): 355-368.

Roberts, J. and Scapens, R. (1985), 'Accounting Systems and Systems of Accountability: Understanding Accounting Practices in the Organizational Contexts', in *Accounting, Organizations and Society*, 10(4): 443-456.

Roman, R. and Colle, R.D. (1999), *Communication Centers and Developing Nations: a State-Of-The-Art Report*, Ithaca, NY: Cornell University.

Sen, A. (2001), *Development as Free*dom, Oxford: Oxford University Press.

Subrahmanian, K. K. (1990), 'Development Paradox in Kerala: Analysis of Industrial Stagnation', in *Economic and Political Weekly*, 25(37): 2053-2058.

Tharamangalam, J. (1998), 'The Perils of Social Development Without Economic Growth: the Development Debacle of Kerala, India', in *Bulletin of Concerned Asian Scholars*, 30(1): 23-34.

Venugopal, V. and Yilmaz, S. (2009). Decentralization in Kerala: Panchayat Government Discretion and Accountability, in *Public Administration and Development*, 29(3): 316-329.

Véron, R. (2001), 'The 'New' Kerala Model: Lessons for Sustainable Development', in *World Development*, 29(4): 601-617.

Chapter 15
Sustainable Broadband Connectivity Model for Rural Areas of Tanzania

Fatuma Simba, Lena Trojer & Zaipuna O. Yonah

1. Introduction

Rural areas are often characterized by highly sparse population consisting of low income communities, where individuals can rarely afford the costs of ICT services (Siochrú, 2005). Therefore, shared access is the model commonly used to deliver ICT services in rural areas. Implementation of the shared access ICT facilities for rural areas differs widely ranging from donor funded projects, in which the project management remains under the local community at the end of the donor funding time, franchise business model, Public-Private Partnership (PPP) and Public-Private People's Partnership (PPPP). Brief descriptions of the mentioned models are given in the forthcoming paragraphs.

A donor funded project is a project in which almost all resources required for project implementation are provided by the donor. Most of the donor funded projects are of the form of pilot projects. They often use a "top-down" implementation approach, of which at the end of the donor funding time, the local community may not have enough capacity or knowledge to manage and maintain the project. Therefore, most of the donor funded projects are usually unsustainable after the donor funding time elapsed. A franchise is a business system in which private entrepreneurs purchase the rights to open and run a similar or part of a big business. Nokia Siemens Network (NSN) village connection, n-logue's rural connectivity model and Grameens public payphone are examples of franchising models implementing ICT projects in rural areas. Characteristics of a franchising model are to push delivery and management of services closer to the end users. The NSN village connection is a solution that enables the Global System for Mobile communication (GSM) network operators to extend their network reach to accommodate remote villages, where traditional technology and commercial business models would not be profitable. The NSN village connection is implemented in partnership with local entrepreneurs, who are familiar with potential customers and who can run the local operations. In this model the entire village customer acquisition and care are handled locally, reducing overall sales and marketing expenses to the operator (franchisor). The n-logue´s rural connectivity model employed a three-tier franchise model to implement a village-level kiosk. The locally owned franchises offer a variety of Internet and computer-based services to the rural market. Partners of this model further developed relevant local content and services using the local language aimed at making kiosks useful to the rural society (Paul, 2004 ; Jhunjhunwala *et al.*, 2004). Another example of a franchise model for rural ICT is the Grameen public payphone. The Grameen public payphone aimed at connecting rural areas of Bangladesh through the provision of public shared-access, where the entrepreneurs provide telephone services to local communities for a fee. The project is implemented by a partnership of the GSM network operator and stationary phones in the villages owned and operated by local entrepreneurs (Cohen, 2001). The franchise model is not widely deployed in implementing ICT projects due to the low level of ICT literacy and low income of the rural community. It is rare to find an entrepreneur with both the skills and financial capacity to operate ICT projects in rural areas. It is obvious that providing connectivity to rural areas is not financially viable to commercial operators. Therefore, government intervention is required for rural connectivity viability. This requirement scenario resulted into the Public-Private Partnership (PPP) model gaining popularity, especially in developing countries, in a situation when a government prioritized rural connectivity.

Public-Private Partnership is collaboration between the public sector such as the government and the private sector to implement projects. In this type of partnership, governments have been intervening by providing financial support in terms of subsidies, often from universal access funds initiatives. While PPPs have the potential to run investments in rural areas, it has been experienced that many developing countries do not have enough administrative and regulatory capacities to run PPP projects efficiently. A typical example is vandalization or poor maintenance of equipments especially those intended for ICT projects. This problem is also partly attributed to lack of beneficiary sense of ownership and inclusion (Spio-garbrah, 2010). This shortcoming was taken as an opportunity to involve beneficiaries of the project in the PPP model, which eventually lead to the Public-Private Peoples Partnership (PPPP) model, sometimes abbreviated as 4Ps.

Public-Private Peoples Partnership is defined as a strategic cooperative relationship between the public sector, private sector and the community (people) who agree to share the responsibility for achieving some specific goals (Siochrú & Girard, 2005; Spio-garbrah, 2010). The partnership emphasizes greater involvement of the communities that will eventually benefit from the outcome of the cooperative relationship. The PPPP model is based on the concept that for projects – especially in the rural areas – to be sustainable, the beneficiaries must feel a sense of ownership and inclusion (Spio-garbrah, 2010). In a PPPP, the ownership of the activity under cooperation such as a project is shared. The heart of a PPPP is therefore sharing of risks and benefits. ICT projects that are implemented in collaboration with the community they serve are usually sustainable. This is evident in the winning projects which were selected as the most successful community-based rural ICT projects in the final report of the Commonwealth African Rural Connectivity Initiative (COMARCI). All of the selected projects were developed in partnership with the community they serve (Calindi *et al.*, 2008; Spio-garbrah, 2010). Therefore, such a selection supports the recommendation that a PPPP is a suitable formula for implementation of projects that are unlikely to be profitable in rural areas if owned and managed either by the public sector alone, the private sector alone or a combination of public and private sector partners (Spio-garbrah, 2010). The schematic diagram of the PPPP model is shown in Figure 1.

Figure 1: Schematic Diagram of Public-Private Peoples Partnership (PPPP)

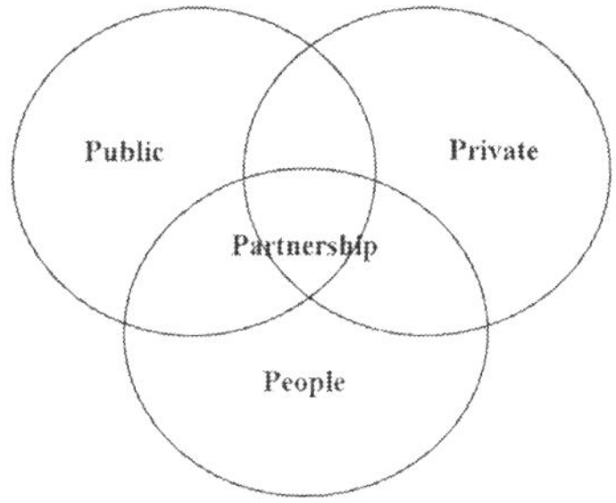

The rest of this chapter is organized as follows: Section 2 presents an overview on how to implement sustainable projects. Three case studies of best practices for sustainable project implementation based on PPPP model are presented in order to understand role of different actors and how collaboration/partnership takes place. Section 3 examines the readiness of Tanzania to provide broadband rural connectivity in terms of availability of ICT infrastructure and institutional frameworks (policies, legislations (Acts) and regulations) supporting rural connectivity in the country. With the existing ICT infrastructures, institutional frameworks and experience from selected best practice projects implementation, a sustainable broadband rural connectivity model adapting the Public Private People's Partnership (PPPP or 4Ps) implementation strategy for Tanzanian context is proposed in section 4. Section 5 carries the conclusion.

2. Sustainable Implementation of ICT Project in Rural Areas

The challenging environments of rural areas result into lack or limited ICT infrastructures, especially the broadband connectivity. In addition, the common models (donor-funded, franchising and PPP) used to implement ICT-based projects in rural areas have shortcomings that lead to unsustainable ICT projects, as explained in section 1. From the literature, the Public Private Peoples Partnership (PPPP) is recommended as a sustainable model for implementation of ICT-based projects in rural areas (Spio-garbrah, 2010; Calindi *et al.*, 2008; Roman, 2009). Also, it has been noted that partnership with the community to be served is essential for the project's success and sustainability (Calindi *et al.*, 2008; Roman, 2009). However, the PPPP concept and its implementation is still in an infant stage (Spio-garbrah, 2010). Therefore, this study aims at adapting the PPPP model and proposing its implementation strategy in the Tanzanian context with the main goal of attaining sustainable broadband connectivity in the rural areas of Tanzania. The proposed PPPP model implementation strategy for broadband connectivity in Tanzania can be generalized to other developing countries, because their respective rural areas usually pose similar characteristics with regard to ICT infrastructure development. The following three projects: rural wireless broadband networks in Peru, rural broadband connectivity project in Dominican Republic and sanitary intervention project in Nigeria, were selected as examples of best practices in implementation of the PPPP model, in order to understand the role of different actors and how collaboration/partnership takes place.

The Rural Wireless Broadband Network in Chancay-Huaral Valley District in Peru

The Chancay-Huaral Valley is a coastal district of Peru with many small farms. Farmers of Huaral were unable to capitalize on the great economic potential of their fertile land, abundant water and proximity to the nearby markets within Peru. A major challenge has been farmers' inability to adjust agricultural production to market fluctuations due to poor communications. Poor communications has also resulted into inefficient management of the river waters for irrigation (Saravia, 2005).

A rural wireless broadband network was established to make agricultural information available to farmers. The specific objectives were: (1) To train farmers in the effective use of agricultural information, (2) To develop an agricultural information system that efficiently provides information via the internet to the farmers, (3) To strengthen local capacity for obtaining, distributing and using agricultural information (Saravia, 2005). The project deployed a wireless network that connects 13 telecenters. The project has been deployed with active participation of the local farmers' association, responsible for management of water and irrigation. Technical assistance has been provided by a Non-Governmental Organization (NGO) called CEPES (the Peruvian Center of Social Studies / the Centro Peruano de Estudios Sociales). The NGO provided ICT training to young people who later took ICT jobs in the project (Saravia, 2005). A remarkable strength that provides sustainability in this project is the collaboration climate that has been created. Existing institutions in the valley pull together their forces and resources to plan and execute this project. Services offered in this project were the agricultural information via internet and Voice over Internet Protocol (VoIP). The VoIP service was not a target in the initial project planning; however, it later came up as a specific demand from project beneficiaries. This shows that the local community was pro-active in maximizing the project capabilities.

Managementof the project was divided into two parts: management of the services that the project is providing and management of the equipment and the physical aspects of the network. Service management was taken care of by administrators of the thirteen communication centres, who have been trained for that. On the other hand, the CEPES technical team handled management of the equipment, configuration, maintenance and technical support. Twelve local people were trained to use the network, to identify connection errors, routine maintenance, to manage, configure and administer the network. Upon completion of their training, they were expected to take charge of network administration tasks from CEPES. There was also a plan for the project´s economic sustainability. It was planned that revenue from the internet and telephone services will be used to cover costs of interconnection with the public network and for paying technicians responsible for network maintenance and technical supports in each location.

Rural Broadband Connectivity Project in Dominican Republic

INDOTEL, the Dominican Republic communications regulatory authority launched the rural broadband connectivity projectas part of the government's e-Dominicanastrategy in 2007. The project objectives were to provide telephone services through public call centers and broadband internet service through internet cafes to 508 localities throughout the country (Roman, 2009). This project also aimed at providing training to both end users and entrepreneurs in the use of the telephones and computers as well as in managing public call centres and cyber cafés. The project included a program to develop web pages with important information about each community, such as location, how to reach it, tourist attractions and products made by the local community. In order to promote entrepreneurship, the Internet access and public call centres were encouraged to be owned and managed by local entrepreneurs.

The project implementation started by establishing the status of telecommunications infrastructure in the area and consultations were made to all stakeholders in order to secure their support and involvement in the project. It was established that there were existing small entrepreneurs who use satellite and wireless technologies to install local telecenters, cyber cafes and public call centres, where the local population pay to use the services. The consultation exercise involved stakeholders such as public institutions and local entities, telecommunication operators, service providers and NGOs.

Establishing contacts with civil society agencies and NGOs was a strategic move, since most of them consider internet access as a working tool and would use it to communicate with their distant offices or staff located in rural areas. For this reason, rural broadband connectivity projects are attractive to civil society agencies and NGOs. Before launching the full scale project, the regulator did a pilot implementation in a small community to confirm whether the rural broadband connectivity project was economically and technically viable (Roman, 2009).

In the Dominican Republic, the rural broadband project was financed by using the Universal Access Funds (UAF) through tender processes supervised by the regulator. In this process, potential investors (telecommunication operator companies) are given a certain time to prepare their bids which are evaluated according to tender procedures. The company with a bid that meets all the technical specifications and asks for the least amount of subsidy from the fund is awarded the project to carry out commercial network roll out (Roman, 2009). In this case study, the regulator played a vital role in ensuring provision of rural connectivity.

Sanitary Intervention Project in Nigeria

This was a project aimed at finding a long-term solution to finance a sanitary intervention project. The mission was to promote safe sanitary and hygiene practices through hygiene education and provision of public convenience facilities that are modern, functional and well maintained (Asabia, 2009). The target areas for provision of sanitary convenience facilities were motor parks, markets, open spaces, central business areas, commuter points and high-density areas. The model adopted to finance sanitary intervention was called public-social private partnership (Asabia, 2009). This is basically a 4Ps model where the unmentioned people´s P is represented by the social component of the model. Roles and responsibilities of partnership in the sanitary intervention project were mentioned as follows(Asabia, 2009): State and local government were responsible toprovide sites to locate the public convenience facilities and to provide an enabling environment for project implementation. The facilities construction costs were financed by the private sector; an NGO. The NGO took lead in engaging local community and other stakeholders located in the sites earmarked for construction of public conveniences to ensure support and ownership of the project. Community and other stakeholders were educated about basic health and hygiene requirements and the link between diseases and unsafe sanitary practices with the ultimate aim of bringing about behavioural change. To ensure project sustainability, the NGO runs and maintains the facility for an interim period of two years, after which a memorandum of understanding is signed, to enable the community or stakeholders in the target area to take over the management and ownership of the facility through a governing council comprising of key stakeholder representatives. The facilities are operated on a pay per use system. Revenue generated from usage is used to pay cleaning staff wages and to cover maintenance costs.

There are four (4) lessons that can be drawn from the projects presented as case studies of the PPPP model for project implementation. The first lesson learned is that: the presence of a governmental can make a project move fast as in the case of the Dominican Republic, where the regulator took a lead in all the steps required for a rural broadband project to take off viably. However, it is not a rule of thumb that the government/regulators are the ones to lead the PPPP initiatives, any of the partnership stakeholders can lead, especially the people (local community or local entrepreneurs), because, they are the ones who properly know their needs. This was very evident in Peru´s and Nigeria's projects. Another lesson learned is the importance of local community participation in all project phases in order to deliver relevant and useful services. This aspect is more evident in the case study from Peru, where the local community was able to demand relevant services. On the other hand, all case study projects demonstrated the usefulness of nurturing project ownership by the local communities/local entrepreneurs for sustainability.

3. Readiness of Tanzania to Deploy PPPP Model for Broadband Rural Connectivity

The following section investigates the availability of supporting environments for rural connectivity, such as the availability of telecommunication infrastructures, institutional framework (policies, regulations, strategies and legislations) and initiatives for infrastructure development, which can be used to gauge readiness of Tanzania to deploy PPPP model for broadband rural connectivity.

Telecommunication Infrastructures

The Tanzania Communication Regulatory Authority (TCRA) is responsible for licensing and enforcing license conditions of broadcasting, postal, and telecommunication operators, manage radio frequency spectrum and establish standards for regulated goods and services. In addition to the mentioned responsibilities, TCRA requires all operators to expand their services up to rural areas in their license obligations. By the year 2011, TCRA had licensed two wired operators, Tanzania Telecommunication Company Limited (TTCL) and Zanzibar Telecommunication Company Limited (Zantel) to provide fixed-line telecommunication services, network services and facilities throughout the country (Tanzania mainland and Zanzibar). The wireless mobile communication technology entered the Tanzanian market in 1994, when the then Mobitel Company, currently branded as Tigo, was first registered as the only privately owned wireless mobile network operator. Currently there are seven wireless mobile operators, namely Tigo, Vodacom, Zain, TTCL, Zantel, Benson and Sasatel. Their respective quarterly (Jan -March, 2011) subscription market share is shown in Figure 2.

Network coverage of fixed-line network operators is mainly up to levels of regional or district headquarters. Apart from the wired network, the wireless infrastructures are also in place from the mobile wireless cellular networks operators, which provide voice as well as broadband data and internet (multimedia) services. With wireless operators, coverage for voice services is almost countrywide while for broadband multimedia services, the coverage is in the urban areas, basically in cities. Furthermore, the TCRA which is responsible for allocating radio frequency to telecommunication networks and ICT services operators, has strategically reserved the extended GSM spectrum for operators aiming to deploy broadband network coverage in suburban and rural areas of Tanzania. So far there is an operator, called Rural NetCO, which has been allocated the extended GSM spectrum and is dedicated to extending coverage of broadband mobile wireless networks to rural areas in the country. This network is based on the third generation (3G) Universal Mobile Telecommunication Systems (UMTS) at 900MHz frequency band. It is a shared network, in which its design strategy is to re-use existing infrastructures such as sharing network base station sites with other telecommunication operators. During the time of writing this chapter, the network was still under development. Therefore, with the current status of both wired and wireless technologies, rural areas of Tanzania are still unconnected with regards to broadband network capacity.

Figure 2: Quarterly Operator's Subscriptions Market Shares (January-March, 2011)

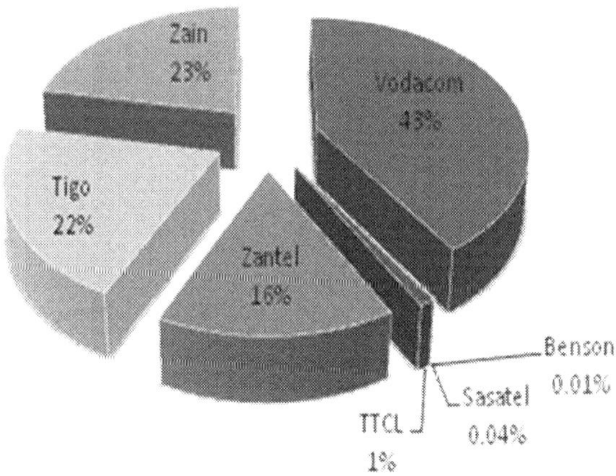

Initiatives for Broadband Infrastructure Development in Tanzania

There are efforts in Tanzania that aim to develop broadband ICT infrastructures. A typical example of such efforts is the National ICT Broadband backbone (NICTBB) infrastructure project. This is a construction of a terrestrial national-wide ICT backbone by using optic fibre cable technology to cover all regions and district headquarters in Tanzania mainland and Zanzibar. The implementation strategy of the national ICT backbone project is to use the existing networks segments from multiple utility companies such as Tanzania Electric Supply Company (TANESCO), Tanzania Railways Corporation (TRC), Songo Songo Gas Company (SONGAS), Tanzania Zambia Railways Authority (TAZARA) and Tanzania Telecommunication Company Limited (TTCL) as shown in Figure 3. The national ICT backbone aims to establish a carrier of carriers' network, which can be leased to network services or content providers. In addition, the backbone is aimed to provide cross border connectivity to the neighbouring countries: Kenya, Uganda, Rwanda, Burundi, Democratic republic of Congo (DRC), Zambia, Malawi and Mozambique. Phase 1 of the NICTBB project which includes utilization of the existing optic fibre cable over pylons owned by TANESCO and construction of new direct buried fibre system along routes of missing links, was completed in April, 2010.

Figure 3: The National ICT Backbone in Tanzania

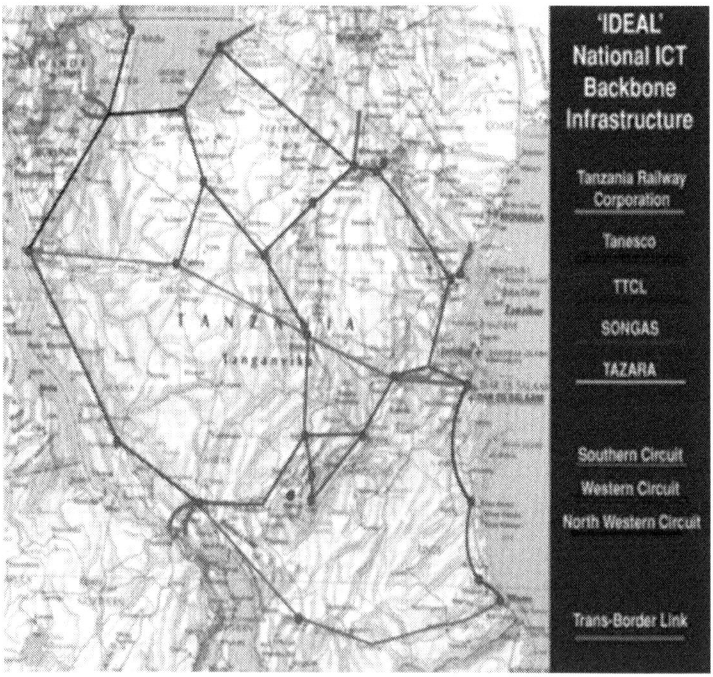

*Source:*Pehrson & Ngwira (2006)

Tanzania is also connected to the rest of the world by using submarine cables, in addition to satellite systems. Examples of such submarine cables are the SEACOM and the Eastern Africa Submarine Cable System (EASSY) along the east African coast. The SEACOM cable was launched on 23rd July 2009 and has a capacity of 1.28Tb/s (terabits per second). EASSY has a capacity of 1.4 Tbps and was launched on 7th April, 2010. The current status (2012) of broadband infrastructure, both terrestrial (NICTBB) and submarine cables in Tanzania is shown in Figure 4.

Figure 4: Tanzania national ICT backbone and Submarine Cables

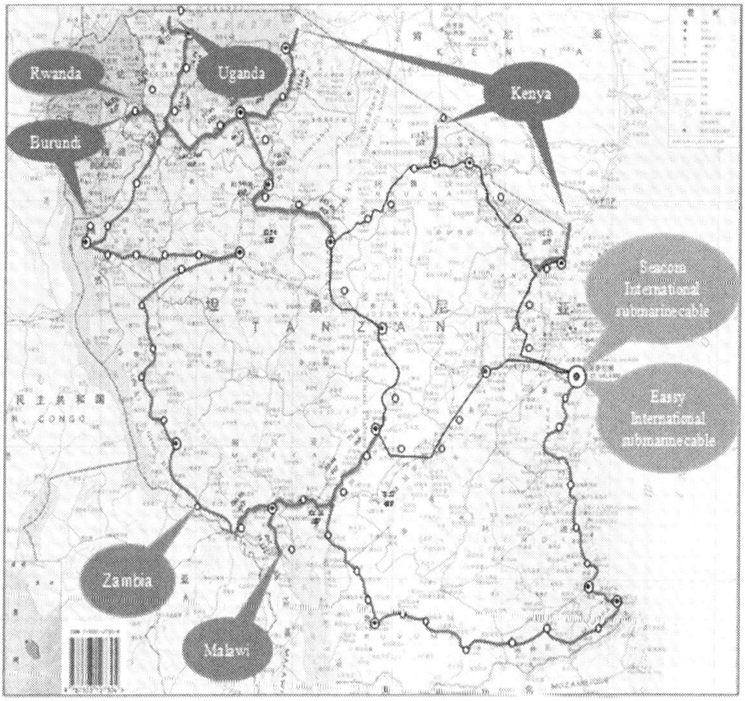

Source: Ministry of Communications, Science and Technology (MCST), Tanzania.

Institutional Frameworks

This section explores institutional frameworks such as policies, regulations, legislations and strategies which support rural connectivity in Tanzania. Existence of such frameworks is a useful tool to enforce and foster implementation of rural connectivity projects.

a. Tanzania National Development Vision 2025

The national development vision aims at a Tanzania of 2025 to be a nation imbued with five main attributes: (1) High quality livelihood (2) Peace, stability and unity (3) Good governance (4) A well educated and learning society (5) A competitive economy capable of producing sustainable growth and shared benefits. The objective of this development vision is to awaken, co-ordinate and direct people's efforts, minds and national resources towards the five main attributes. Upon achievement of the five attributes, Tanzania is envisioned to be an active participant in the global competitive economy of the 21st century, which is characterized by advanced technology, high productivity, modern and efficient transport and communication infrastructure with highly skilled and innovative manpower (United Republic of Tanzania, 2000).

b. National ICT Policy

In 2002, Tanzania set off a multi-stakeholder ICT policy development initiative that led to the development of a national ICT policy and its adoption by the Government in 2003. The policy is currently (2011) under review for improvement. The policy articulates ten focus areas drawn from the aspirations of Tanzania's Vision 2025. One of the focus areas is the ICT infrastructure, with an objective of establishing mechanisms that will result in least cost access to bandwidth for institutions or individuals in Tanzania. The National ICT Policy stipulates that, "Tanzania should have a universally accessible broadband infrastructure and ICT solutions that enhance sustainable socio-economic development and accelerate poverty reduction national-wide" (Ministry of Communications and Transport, 2003). To achieve this objective, the Government is committed to support, through incentives and directives, bona fide institutions actively involved in the development and application of ICT infrastructure and services (Ministry of Communication and Transport, 2003). Another focus area of ICT policy is the universal access, where the Government is committed to reduce the ICT access gap between the rural and the urban areas by activating the universal communications access fund, offering special incentives to investors in rural ICT provisions, supporting the construction of rural telecentres and involving local government authorities in ICT utilization and promotion (Ministry of Communications and Transport, 2003). As part of the implementation of the national ICT policy, Tanzania embarked on an initiative for development of an open national ICT backbone network in 2005.

c. The National Strategy for Growth and Reduction of Poverty (NSGRP)

Introduced in 2005, this strategy which is popularly known as MKUKUTA in Swahili, was revised in 2010 into MKUKUTA II. Among other objectives, the strategy aims to promote the use of PPP model for developing and operating ICT infrastructures. It also looks for efforts to support increased access and application of ICT which will eventually increase productivity in government, business, education and Small To Medium Enterprises (SMEs) development (NSGRP, 2005; NSGRP II, 2010).

d. National Rural Development Strategy (RDS)

The strategy came into existence in the year 2001 with the aim to promote introduction of ICT in rural areas through developing telecentres. The telecentres are responsible for providing ICT services such as voice, fax, e-mail, internet access, distance education, telemedicine, news distribution, access to information on market and weather conditions (RDS, 2001).

e. Universal Communications Service Act 2006

The Act established a fund responsible for promoting ICTs in rural and underserved urban areas by subsidizing operators investing in rural areas (UCS-Act Tz, 2006; Pehrson & Ngwira, 2006). The fund receives contributions from TCRA, development partners and telecommunication operators who contribute 0.3% from their annual growth turn over.

f. Converged Regulatory and Licensing Regime

In 2005, TCRA introduced a converged licensing framework, which is a technology- and service - neutral framework. The converged framework ensures efficient utilization of network resources, in such a way that an individual network can be used to provide a broad range of ICT services, which will eventually lead to address market and technological developments. The converged licensing regime allows leasing of excess capacity of communications infrastructures owned by utility companies in order to provide communication services to customers. Moreover, it provides a framework for all communication service providers to build a nationwide backhaul and interconnect with each other.

In 2005, Tanzania prepared different communication regulations which in one way or another support provision of broadband connectivity. Regulations relevant to broadband connectivity are the Tanzania communication regulations for: broadband services, quality of service, access and facilities, radio communication and frequency spectrum, and interconnection.

Table 1: ICT Initiatives for Education

Initiative (projects)	Description
Tanzania beyond Tomorrow	An e-learning initiative launched by MoEVT in partnership with technology companies (Accenture, Intel, Microsoft and Cisco) and NGOs (NetHope, AMREF, World Vision) aims at integrating ICT in the teaching and learning processes.
TanSSe-L (CoICT-UDSM)	The project is under the College of Information and Communication Technology (CoICT), University of Dar es Salaam. It aims at providing e-learning services to rural secondary schools. It has three pillars: interactive learning management system, e-learning contents for self learning and connectivity solution.
The e-School Forum	Nine task forces were formed, focusing on key areas for ICT in secondary education (1) ICT infrastructure development (2) technical resources, (3) student management (4) content and curriculum development (5) e-learning (6) sensitization (7) human resources (8) programme co-ordination and (9) funding. At the time of writing this chapter, the e-School programme proposal was still under review by the MoEVT.
Science, Technology and Higher Education Program (STHEP).	It is a programme under the MoEVT and the Ministry of Communication Science and Technology (MCST) that aims at improving development of human capital in the areas of science and technology and to create a knowledge based economy. This program focuses on National Research and Education Network (NREN), Education Management Information system (EMIS), e-Library and e-Learning.
Zain, Build a Nation project	670 schools in Tanzania have benefited with textbooks or educational tools such as computers provided by the then Zain, now Airtel mobile network operator.
Bridge IT project	The project is implemented under the management of International Youth Foundation. It is about the use of mobile phones to teach Maths and Science in class 5 and 6. It is operational in 150 primary schools. Development of course materials is done by Pearson Foundation, hosting of the course materials is by Vodacom. Delivery of the course materials is through Nokia phones to television sets in the classrooms. Contents assist teachers in the classrooms.
Computer for Schools Project	Barclays bank and Digital Links International company formed a partnership to spur the growth of ICT in schools across three countries in East Africa (Tanzania, Uganda, and Kenya). The project sets a target to place 10,000 computers in approximately 500 schools. The implementation strategy is through collaborative partnerships with organizations in each country. For Tanzania, the project is implemented by Tanzania Education Authority (TEA), Tanzania Commission of Science and Technology (COSTECH), and Mkombozi centre for street children.
Distance Learning Education Services (DILES) (IICD Supported)	The project develops teaching and learning materials for secondary schools. These resources are available online (http://www.diles.or.tz).

g. ICT for Education Projects

In Tanzania, there are number of pilot projects aimed at integrating ICT in the education system. The projects are currently challenged by lack of connectivity to make them more beneficial. Therefore, most of the existing ICT for education initiatives aim at providing computers to schools, and sometimes training to teachers and students. The existing projects on ICT for education are formed under different collaboration models; the most common model being a Public-Private Partnership (PPP), which is a collaboration between the Ministry of Education and Vocational Training (MoEVT) with technology companies. Another model is the donor funded model, where the efforts are from either the private or public sector to support schools by providing computers or capacity building through training. Table 1 provides a brief description of indicative sample projects. The list is not exhaustive.

The ICT for education initiatives summarized in Table 1, illustrates that Tanzania has a number of ICT for education initiatives running concurrently. Many of these initiatives are pilot projects that are not well documented with no clear strategic plan for their implementation and there is no centralized coordination. The missing coordination and strategy for implementation to the many ICT for education initiatives implies that opportunities to combine, or associate related projects are missed, hence duplication, lack of experience sharing, and a potential for ineffective use of resources and opportunities is inevitable. Tanzania is currently (2011-2012) preparing a National School Connectivity Plan (NSCP) under the Ministry of communication, science and technology in collaboration with International Telecommunication Union (ITU). The NSCP aims at providing an ICT broadband infrastructure master plan to facilitate implementation of broadband connectivity solution for primary and secondary schools as well as to tertiary education. Among other targets, the NSCP intends to address the lack of coordination and absence of a single ICT for education implementation strategy.

This section presented the existence of telecommunication infrastructures, institutional frameworks and initiatives for broadband connectivity in the country and ICT for education projects. Collectively, all these efforts demonstrate readiness of Tanzania to provide rural connectivity. The following section proposes implementation strategies in order to build sustainable broadband rural connectivity in Tanzania.

4. The Proposed Model for Provision of Sustainable Broadband Rural Connectivity

Availability of broadband infrastructures, institutional frameworks, ICT for education projects and the governmental will are the potential enabling tools for PPPP implementation in Tanzania. Having the tools and knowledge from the best practice projects implemented based on PPPP model, a sustainable model for broadband rural connectivity suitable for the Tanzanian context is proposed as shown in Figure 5. Roles of the different actors in the model are explained in the forthcoming sections.

Public Sector

The public sector can be made up of the communication regulatory authority and/or the Ministry responsible for communications, science and technology. The main roles of the public sector are:

- To create and enforce enabling policy and regulatory frameworks for PPPP.
- To train the local community in basic computer literacy, Internet use as well as services and applications delivered by broadband networks.
- To enforce as well as contribute the subsidy funding for universal communications access fund to private sectors in order to facilitate ICT infrastructure and service deployment to rural areas.
- To encourage the private sector to rollout commercial networks in rural areas.
- To coordinate formation of the PPPP.

Figure 5: The PPPP Model of Rural Broadband Connectivity in Tanzania

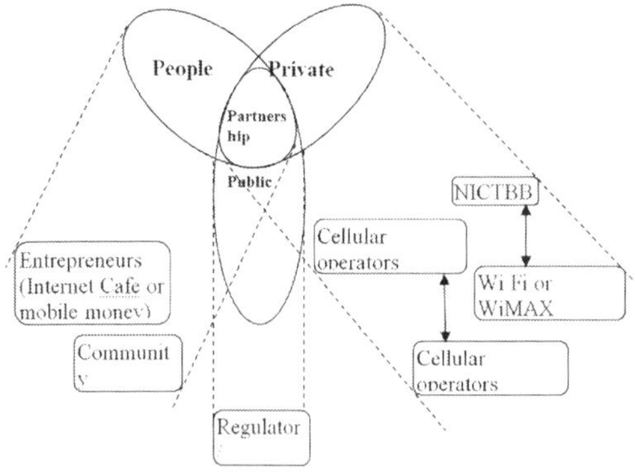

Private Sector

The private sector in this context is proposed to have two (2) tiers; the upper tier also known as the backbone tier and the lower tier known as the access network tier. The architecture of the private sector in the Tanzanian PPPP model is shown in Figure 6. The upper tier (backbone tier) is composed of cellular mobile operators. These are the operators that have wide coverage in the urban areas, but their networks are not extended to reach rural areas, due to the high cost associated with network roll out and maintenance in rural areas. Another candidate for the backbone tier is an operator managing the NICTBB infrastructure. The objective of the backbone tier is to provide local and international connections to the access network tier, which is made up of operators who have infrastructures in rural areas. The access network tier is responsible to provide broadband network capacity in the rural areas. In this context, the access network tier can be made up of Wireless Fidelity (WiFi) or Worldwide Interoperability for Microwave Access (WiMAX) wireless operators extending the coverage of the NICTBB. The access network tier can as well be made up of a wireless cellular mobile operator of 3G broadband access network at 900MHz frequency band to extend coverage of the existing 3G at 2100MHz broadband mobile wireless network.

Figure 6: Two Tiers of Private Sector in the PPPP Model for rural connectivity in Tanzania

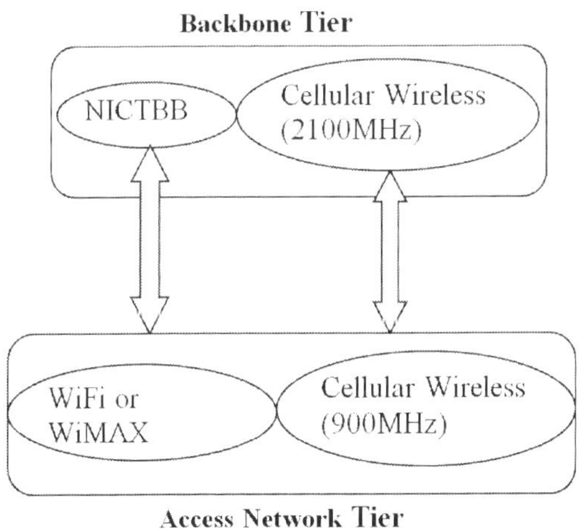

Different roles of the two tiers are as follows:

Backbone Tier

- Provide local and international connections to the access network tier
- Develop and offer value added services
- Create environment for franchise business model with access network tier
- Train the access network operators for potential business opportunities
- Train the local community in basic computer literacy, Internet use as well as services and applications delivered by broadband networks.

Access Network Tier

- Roll out the commercial broadband access network infrastructure
- Develop and offer value added services
- Create environment for franchise business model with local entrepreneurs
- Train local entrepreneurs for potential business opportunities
- Train the local community in basic computer literacy, Internet use as well as services and applications delivered by broadband networks.

People´s Sector

The people or local community can be the local government or management responsible for a shared local ICT facility such as telecentres. Another potential local partners to represent the people's part are the existing entrepreneurs in rural communities who are already providing some forms of telecommunication services such as internet cafes or mobile money transfer to others within the community. The PPPP model can benefit from the ICT-based local entrepreneurs from their ICT literacy and experiences. Roles of the people's part in the partnership can be:

- Make in-kind contribution to set - up network, for example provide land, labour and site security.
- Train the local community in computers and Internet use.
- Strive for participation and local ownership of the rural broadband connectivity, which will lead to sustainability and service relevance.
- Demand and maximize use of the value added services.

Sustainability aspects of the proposed PPPP model for broadband rural connectivity in Tanzania are considered into two aspects: financial and the partnership (stakeholders working together) sustainability. In addition to the subsidies offered by the public sector and revenue collected from the main intended service to be offered, other value added services relevant to the local community can contribute to the rural connectivity project´s financial sustainability. Providing need-driven services that are affordable to the local community, which also inspire social-economic development at the community level will generate revenue that in turn contribute to the financial sustainability of the project.

The training component appears to be a role for all actors in the proposed PPPP model due to the fact that capacity building to the local community is essential to ensure sustainability of the project. People in the community have to understand the usefulness and potential of broadband Internet as well as of the services and applications that it can deliver. If the local community feels empowered by the broadband rural connectivity projects, they will participate more actively in seeking ways to keep it running. Capacity building by formal training is not the only ingredient for sustainability. Awareness creation to the community about services to be provided by the broadband network and how to access and use them can also add to the sustainability of the rural broadband connectivity. Therefore, it is necessary for the people to be trained to become customers of the service and thereby ensure the project's sustainability. Training for potential business opportunities between backbone tier, access network tier and local entrepreneurs is essential in order to foster business partnership climate. In order to enhance sustainability, the partnership should strive to nurture local community inclusion, involvement and ownership throughout all the implementation phases of rural broadband connectivity projects.

Participatory action research is a practical approach which involves all participants affected by the issue under study (Rydhagen, 2002). Active participation of actors, which can be brought about by the participatory approach, strengthens partnership and trust among the stakeholders. Due to the requirement of PPPP model to maintain the strategic partnership in all phases and throughout the project, the participatory approach is recommended in the process of nurturing partnerships, because it can build a climate where rural communities feel a sense of inclusion and ownership. Furthermore, having the existing policies, legislations, strategies and regulatory framework supporting rural connectivity will protect, promote and support its development and existence, hence its sustainability.

5. Conclusion

This study has identified the availability of supporting and enabling tools such as broadband connectivity, Integration of ICT in the educational system, governmental will and institutional frameworks, favourable for adopting PPPP model for rural broadband connectivity in Tanzania. By having such tools and the supporting environments, a PPPP model for sustainable rural broadband connectivity implementation in Tanzania has been proposed. The Government of Tanzania through its Ministry of Communication, Science and Technology in collaboration with International Telecommunication Union (ITU) are formulating the national school connectivity plan (NSCP). The objective of the NSCP is to develop a nation-wide master plan for providing broadband connectivity to all schools in Tanzania. The PPPP model for rural broadband connectivity proposed in this study can be a very useful input to the NSCP.

References

Asabia, A. (2009), 'Presentation on public, private and people partnership model for sanitation at the 6th earth watch conference on water and sanitation 2008', in *International NGO Journal*, 4(8): 362-367.

Calindi, V., Tongia, R., Ross, P., Pulkkinen, J., Akwetey, L. and Field, D. (2008), 'The Commonwealth African Rural Connectivity Report', in *Commonwealth African Rural Connectivity Initiative (COMARCI)*.

Cohen, N. (2001), 'What Works: Grameen Telecom's Village Phones', in *A Digital Dividend Study by the World Resources Institute.*

Jhunjhunwala, A., Ramachandran, A. and Bandyopadhyay, A. (2004), 'n-logue: the story of a rural service provider in India', in *The Journal of Community Informatics*, 1(1): 30-38.

NSGRP (2005), 'The United Republic of Tanzania National Strategy for Growth and Reduction of Poverty (NSGRP)', United Republic of Tanzania, Vice President´s Office, Dar es Salaam.

NSGRP II (2010), 'National Strategy for Growth and Reduction of Poverty II', United Republic of Tanzania, Ministry of Finance and Economic Affairs, Dar es Salaam .

Paul, J. (2004), 'What works: n-Logue´s rural connectivity model: deploying wirelessly-connected internet kiosks in villages throughout India', Washington, DC: World Resources Institute. Available at: http://www.digitaldividend.org/pdf/nlogue.pdf

Pehrson, B. and Ngwira, M. (2006), *Optical Fibre for Education and Research Networks in Eastern and Southern Africa,* University of Malawi. Available: http://www.idrc.ca/uploads/user-S/11083295861PAREN_Reportv15.doc

RDS (2001), 'Rural Development Strategy (RDS)', Prime Minister`s Office, United Republic of Tanzania, Dar es Salaam.

Roman, E. S. (2009), 'Bringing broadband access to rural areas: a step-by-step approach for regulators, policy makers and universal access program administrators: the experience of the Dominican Republic', in *9th Global Symposium for Regulators (GSR),* 10 - 12 November, Beirut, Lebanon.

Rydhagen, B. (2002), 'Feminist Sanitary engineering as a participatory alternative in South Africa and Sweden', Doctoral Dissertation Series Number: 2002:06, Blekinge Institute of Technology.

Saravia, M. (2005), 'Rural telecommunication networks in Peru, case studies 3 and 4 of innovative technologies and community ownership: a new model of ICT access for the rural poor', in *A report for the United Nations Development Programme (UNDP), Series: Making ICT Work for the Poor*, 99-141.

Siochrú, S. Ó. (2005), 'Community ownership of ICTs: new possibilities for poor communities', *WSIS Briefing Paper*, 1-6.

Siochrú, S. Ó. and Girard, B. (2005), 'Community-based networks and innovative technologies: new models to serve and empower the poor', *A report for the United Nations Development Programme (UNDP), Series: Making ICT Work for the Poor*, 185.

Spio-garbrah, E. (2010), 'Driving rural connections: leveraging public-private-people's partnerships to improve rural connectivity and adopting the 4P approach to develop, promote and accelerate ICT connectivity in the developing world', in *Mobile World Congress 2010, Special Report From Barcelona*, (4): 30 - 38.

Ministry of Communications and Transport, The United Republic of Tanzania (2003), 'National Information and Communications Technologies Policy', Available at: http://www.tanzania.go.tz/pdf/ictpolicy.pdf

UCS-ActTz (2006), 'Universal Communication Service Act', United Republic of Tanzania, Vice President´s Office, Dar es Salaam.

United Republic of Tanzania (2000), 'Tanzania National Development Vision 2025'. Available at: http://www.tanzania.go.tz/vission_2025f.html

Acknowledgement

Authors are grateful to the Swedish International Development Agency (Sida) for sponsoring this study.

Chapter 16
Post Scriptum

Mammo Muchie & Angathevar Baskaran

Today ICT is transformative and is creating the innovation commons as shared space and resource for accessing new digital data globally. There has been a flagship report produced in 2012 by the African Development Bank, the African Union in collaboration with the World Bank by identifying key sectors such as –agriculture, climate adaptation, education, financial services, health and Government including local ICT sector and regional trade and integration. How ICT transformation-ready these sectors are, is what the report with the title **eTransform Africa** attempted to address (World Bank & African Development Bank, 2012). The report acknowledged Africa is turning into an ICT global hub producing dual SIM card for cellular mobile phones and M-Pesa for remittance payments. They acknowldged Africa has turned into a global market for mobile subscriptions more than the Western world by reporting that something like over 750 million people are having access to the service.

The eTransform Africa report also acknowledged that ICTs are facilitating and stimulating inter pan-African regional trade and integration. New technology driven industries are also growing creating innovation promoting African youth employment and industrial exports. Job creation, innovation and entrepreneurship are growing and both the private and public sector are applying ICTs to accelerate their productivity and services. The report asserts that bandwidth internet access has grown 20-fold in less than four years time. Something like two thirds of African youth uses mobile phone subscription by the year 2013 when the report was issued. In Kenya, there is a home grown Kilimo salama scheme to provide crop insurance to farmers by using M-PESA mobile payments by dealing with natural hazards related to either drought or huge rainfall. In Malawai communities are trained using ICTs services to address deforestation projects to deal with and respond to climate change hazards by using GPS devices to engage and address by developing localised climate adapation strategies. In Senegal they have financial services by putting forward their own money transfer through an exchange SONTEL scheme that is a subsidiary of Orange. In Mali they have a telemedicine by establishing the IKON Tele-radiology program to address the lack of trained health care specialists in the rural areas.

It is very interesting to see such examples and attempts to make Africa-made solutions by using ICTs with the hope of turning Africa as a knowledge hub. This is indeed a truly very encouraging development. It is very early to say whether these concrete examples can be developed and expanded to cover the whole continent. In both kenya and Rwanda there are developments in ICTs that can be resources for learning for other countries in Africa.

For and in Africa, of all the sectors the most significant is to apply ICTs for transforming agriculture in order to eliminate unemployment, poverty and inequality in the continent. Africa needs food security as a human right. Transforming agricultural with the successful Information and communication technology revolution is one of the most strategic challenges that confronts Africa today. The report is that the agriculture sector is not performing as expected in and for Africa. 73 % of the farmers are still poor in much of Africa, and they are said to live with less than one dollar a day.Advanced technologies and infrastructre are necessary to facilitate the transformation of agriculture. ICTs can bring changes and can fill in the knowldge deficits. There is a need to make ICTs as enablers and knowledge promotors by empowering, enabling and entitling the farmers to perform sustainable agricultural transformation by dealing with all the climate variations and challenges.

There is a need to work to create a sustainable ICT-enabled transformations in Africa by encouraging all stake-holders to incentivise and support all the home-grown best examples that we highlighted in Kenya, Rwanda, Malwai, Mali and Senegal.

The contributions in this book address the issue of how ICT can stimulate the innovation commons and resources to promote the transfrmation and sustainbe development in Africa. Despite the optimistic take on ICTs in Africa, there is still a lot to be learned by applying ICTs in areas that matter like agriculture to transform the African predominantly largely agrarian economy. This book has scholarly contributions to facilitate how transformation through ICTs innovation in this time of the digitial age can make a big difference to bring about structural social, environmental and economic transformation to build Africa's future with its own agency applying technology that is necessary to apply.

Reference

World Bank and African Development Bank (2012), *The Transformational Use of Information and Communication Technologies in Africa.* Available at: http://siteresources.worldbank.org/EXTINFORMATIONANDCOMMUNICATIONANDTECHNOLOGIES/Resources/282822-1346223280837/MainReport.pdf

Notes on Contributors

Mammo Muchie

DST/NRF SARChI Professor (Innovation & Development), Tshwane University of Technology, South Africa; Technology and Management Centre for Development (TMCD), Oxford University, UK; Associate Faculty Professor, Sussex University, UK; University of Gondar(UoG) , Arsi University, & Adama Science and Technology University (ASTU), Ethiopia. Email: MuchieM@tut.ac.za

Angathevar Baskaran

Associate Professor, Department of Development Studies, Faculty of Economics and Administration, University of Malaya, Malaysia; Senior Research Associate of SARChI (Innovation & Development), Tshwane University of Technology, Pretoria, South Africa. Email: baskaran@um.edu.my & anga_bas@yahoo.co.uk

Olu Ajakaiye

African Economic Research Consortium (AERC), Nairobi, Kenya. Email: ajakaiye_olu@yahoo.com; olusanya.akakaiye@aercafrica.org

Samuel Wangwe

Executive Director, Research on Poverty Alleviation (REPOA), Dar es Salaam, Tanzania. Email: swangwe@repoa.or.tz

Christopher Freeman

(Late)Emeritus Professor and Founding Director, The Science Policy Research Unit (SPRU), University of Sussex, Brighton, UK.

Luc Soete

Director, UNU-MERIT and Professor of International Economic Relations, Maastricht University, Maastricht, the Netherlands.
Email: soete@merit.unu.edu

Michael Gastrow

Education, Science and Skills Development Research Unit, Human Sciences Research Council, South Africa. Email: mgastrow@hsrc.ac.za

Saahier Parker

Centre for Science, Technology and Innovation Indicators, Human Sciences Research Council, South Africa. Email: sparker@hsrc.ac.za

Timothy Mwololo Waema

School of Computing and Informatics, University of Nairobi, Kenya. Email: waema@uonbi.ac.ke

Winnie Mitullah

Institute for Development Studies, University of Nairobi, Kenya. Email: wvmitullah@swiftkenya.com

Edith Adera

East African Regional Research Co-ordinator, IDRC, Nairobi.

Obadia Okinda

School of Development Studies, University of KwaZulu-Natal, South Africa: 211559228@ukzn.ac.zaPeter Ulanga

Peter Ulanga

Universal Communications Service Access Fund (UCAF), TTCL Kijitonyama Building, Ali Hassan Mwinyi/ Garden Road, P. O. Box 33114, Dar es Salaam, Tanzania. Email: ulanga@ucaf.go.tz;ulanga@gmail.com

John O. Adeoti

Economic Policy Research Department, Nigerian Institute of Social and Economic Research (NISER), Oyo Road, Ojoo, P.M.B. 5 UIPO,Ibadan, Nige-ria; Email: Adeotij@yahoo.com

Foluso M. Adeyinka

Economic Policy Research Department, Nigerian Institute of Social and Economic Research (NISER), Oyo Road, Ojoo, P.M.B. 5 UIPO,Ibadan, Nige-ria. Email: folu_adeyinka@yahoo.com

Moses O. Ubaru

National Information Technology Development Agency (NITDA), Plot 695 PH Crescent, Abuja, Nigeria. Email: moses_ubaru@yahoo.com

Sunday A. Khan

TrustAfrica, Dakar, Senegal; and Faculty of Economics and Management, University of Yaoundé II, Cameroon.Email: sunkhan68@yahoo.co.uk

Arsène Gidéon Nkama

Faculty of Economics and Management, University of Yaoundé II. Email: agnkama@yahoo.com

Latif Dramani

UFR SES CREFAT Université de Thiès BP: 947 Thiès Sénégal. Email: drama-niarmel@yahoo.fr

Oumy Laye

Agence Nationale de la Statistique et de la Démographie. Email : lay-endiaye78@yahoo.fr

Vijay K. Bhasin

Department of Economics, University of Cape Coast, Cape Coast, Ghana. Email: vbhasin96@yahoo.com

Simon K. Harvey

Department of Finance, University of Ghana Business School, Legon, Ghana. Email: sharvey@ug.edu.gh

Eme Umoeka Fiawoyife

Department of Finance, University of Ghana Business School, Legon, Ghana. Email: eufiawoyife@yahoo.com

Mutawakilu A. Tiamiyu

Africa Regional Centre for Information Science, University of Ibadan, Ni-ge-ria. Email: mutatiamiyu@yahoo.com and ma.tiamiyu@mail.ui.edu.ng

Abiodun S. Bankole

Department of Economics, University of Ibadan, Nigeria.Email: as.bankole@mail.ui.edu.ng and asbanky@yahoo.com

Rosemary O. Agbonlahor

Africa Regional Centre for Information Science, University of Ibadan, Nigeria. Email:ro.agbonlahor@mail.ui.edu.ng andragbonlahor@yahoo.com

Silvia Masiero

Information Systems and Innovation Group, London School of Economics and Political Science, London, United Kingdom. Email: s.masiero@lse.ac.uk.

Fatuma Simba

PhD Scholar, University of Dar es Salaam, Tanzania. Email: fatmasimba@udsm.ac.tz

Lena Trojer

Professor of Information and Communication Technology (ICT) for Devel-opment, Blekinge Institute of Technology, Sweden. Email: lena.trojer@bth.se

Zaipuna O. Yonah

Director of Information and Communication Technology (ICT), Ministry of Communication, Science and Technology, Tanzania. Email: zaipuna.yonah@mst.go.tz; zaipuna.yonah@gmail.com

Index

Symbols

A

B

D

E

F

G

H

I

J

K

L

M

N

O

P

Q

R

S

T

U

V

W

Y

Z